An Orchard Invisible

An Orchard Invisible

A Natural History of Seeds

Jonathan Silvertown

WITH ILLUSTRATIONS BY AMY WHITESIDES

The University of Chicago Press
Chicago and London

Jonathan Silvertown is professor of ecology at the Open University in England. He is the author of *Demons in Eden: The Paradox of Plant Diversity* and *99% Ape*, both published by the University of Chicago Press.

The University of Chicago Press, Chicago 60637
The University of Chicago Press, Ltd., London

Printed in the United States of America

17 16 15 14 13 12 11 10 09   1 2 3 4 5
ISBN-13: 978-0-226-75773-5 (cloth)
ISBN-10: 0-226-75773-0 (cloth)

Library of Congress Cataloging-in-Publication Data

Silvertown, Jonathan W.
An orchard invisible : a natural history of seeds / Jonathan Silvertown ; with illustrations by Amy Whitesides.
p. cm.
Includes bibliographical references and index.
ISBN-13: 978-0-226-75773-5 (cloth : alk. paper)
ISBN-10: 0-226-75773-0 (cloth : alk. paper) 1. Seeds—Popular works. 2. Botany—Popular works. I. Whitesides, Amy. II. Title.
QK661.S558 2009
581.4'67—dc22

2008043555

♾ The paper used in this publication meets the minimum requirements of the American National Standard for Information Sciences—Permanence of Paper for Printed Library Materials, ANSI Z39.48–1992.

For my father and
in memory of my mother,
Alfred & Eva

Contents

An Orchard Invisible

{ 1 }

An Orchard Invisible

SEEDS

A seed hidden in the heart of an apple is an orchard invisible.
WELSH PROVERB

Seeds have a mirrored life, the original in nature and another reflected in literature and the imagination. The Welsh proverb simultaneously expresses both the biological potential of seeds and their metaphorical power. The American philosopher and early conservationist Henry David Thoreau, who was fascinated by seeds and inspired by them, wrote, "I have great faith in a seed. Convince me that you have a seed there, and I am prepared to expect wonders." Who cannot wonder that the largest organism on this planet, the giant redwood *Sequoiadendron giganteum* nicknamed "General Sherman," which weighs roughly the equivalent of a fleet of six Boeing 747-400 Jumbo Jets, germinated more than two thousand years ago from a seed weighing only a six-thousandth of a gram?

This book was once a seed or, to be precise, several. The paper in it is made from wood pulp from northern coniferous forests, grown and replanted from seed; the inks used on its pages and the varnish on its cover contain oils obtained from seeds; but the book grew from another kind of seed also—the seed of an idea. The idea was for a book that would explore the science behind all the familiar things that gardeners, cooks, and everyone else knows about seeds. This book is about *why* seeds have all the wonderful properties that they do, feeding us, flavoring our food, moistening and protecting our skin, and growing into plants that give us fruits, flowers, fiber, pharmaceuticals, poisons, perfume, protection, and pleasure. I hope to persuade you that reading about seeds is yet another way to enjoy them.

Expect wonders. Orchids have seeds as light as dust that are so ill-provisioned that seedlings spend their first few years as parasites on fungi. The biggest seed, by contrast, weighs twenty kilograms and belongs to the double coconut, a castaway palm whose ancestors were stranded on tiny islands in the Seychelles after these fragments of land were cast aside by India in its geological passage toward Asia. Evolution on a crumb from the Indian landmass produced Earth's largest seed.

The story of seeds, in a nutshell, is a tale of evolution bursting with questions. Its earliest episode concerns how and when the first seed plant evolved from its fern-like ancestors, with other episodes about why seeds have dormancy and what makes them germinate; why some seeds are rich in oil and others in starch; why some seeds are big and others small; why some are poisonous and others palatable. Though science has unraveled much about the whys and wherefores of seeds, there are still mysteries to be solved, including the most fundamental question of all: why do plants bother producing seeds in the first place? Why are plants, or animals for that matter, so hooked on sex?

You can follow the story cover-to-cover in a journey that will take you from a seed's first beginnings in sex and pollination through each step of its life till it ends in your coffee cup or on your plate; or you can dip into the story and sample a crumb of gardening, a grain of genetics, a dose of medicine, a packet of commerce, or a morsel of cooking: all in a seed. This is not a long book (who has time for those, these days?), but do not expect an express route from A to Z. Rather, I have followed a meandering path through the story of seeds and, like the root of a seedling, I have branched into particularly fertile territory wherever I have encountered it. So indulge me, and you will learn, as I did while writing, some fascinating if occasionally tangential connections between seeds and unexpected topics such as the witchcraft trials in seventeenth-century Salem (chapter 6, "O Rose, Thou Are Sick!"), Lyme Disease (chapter 8, "Ten Thousand Acorns"), human color vision (chapter 9, "Luscious Clusters of the Vine"), and the evolution of yeast (chapter 15, "John Barleycorn").

Whichever way you read the book, you will find a recurring theme: evolution constantly invents new uses for old devices. This happens because of how evolution works, creeping gradually and without direction from one solution to life's challenges to another. The achievements of the process of evolution are astonishing, and to some people unbelievable, but we didn't get here overnight. The two millennia it takes a seed to produce a tree the size of General Sherman is a mere fleeting instant in the 360-million-year history of seed plants. Laozi, a Chinese philosopher writing at about the time General Sherman was just a six-milligram seed, got it right when he reflected, "To see things in the seed, that is genius."

{ 2 }

First Forms Minute

EVOLUTION

Organic Life beneath the shoreless waves
Was born and nurs'd in Ocean's pearly caves;
First forms minute, unseen by spheric glass,
Move on the mud, or pierce the watery mass;
These, as successive generations bloom,
New powers acquire, and larger limbs assume;
Whence countless groups of vegetation spring,
And breathing realms of fin, and feet, and wing.
ERASMUS DARWIN, FROM *The Temple of Nature*

Charles Darwin's visionary grandfather Erasmus was ridiculed by his contemporaries for his evolutionary views. The family crest that he designed and blazoned on his carriage boldly declared "ex omnia conchis," or "everything from seashells." It may have been Erasmus's little joke, but he was decades ahead of his time and fundamentally right: life did first evolve in the sea. Where, though, did seed plants originate? There are indeed some seed plants that live in the sea, but the seagrasses (as they are known) live in shallow coastal waters and have terrestrial ancestors. Seagrasses are mere parvenus to a marine existence, still paddling in the muddy shallows where they can avoid the heavy-hitters among marine plants—the algae.

Even though seed plants first evolved on land, that does not mean that we can ignore the marine origins of land plants themselves. Quite to the contrary—although evolution took the plant out of the sea it couldn't so easily take the sea out of the plant. Or, as E. H. Corner put it in his classic book *The Life of Plants*, land plants "are made from a sea recipe." From a sea recipe, evolution cooked up something brand-new to serve the demands of life on land: an embryo in a box, which we call a seed. In fact the box contains not just an embryo, but also a food store placed there by mother, so a seed is really an embryo in a picnic basket.

Seeds were the ultimate refinement of plant adaptation to terrestrial life. So what preceded them and how did they evolve from a sea recipe? A comparison between plants and animals is illuminating. Among animals, successful colonization of the land from the sea was achieved several times independently by vertebrates, molluscs, and arthropods (insects and crustacea), but among plants there was only one, solitary successful colonization. All land plants, which include mosses, ferns, horsetails, gymnosperms (conifers, cycads, and related groups), and flowering plants, descend from a single ancestor that first made the transition from sea to land. There must have been failed attempts too, but we do not know how many.

That the transition from sea to land was successful only once among plants is no doubt testament to the many obstacles to survival and reproduction that life on land would have presented to a marine alga. Indeed, the differences between terrestrial and marine environments that affect plants are so numerous that E. H. Corner declined to list them in his book, saying "A list would cover many pages and the active mind must be spared the tedium." A praiseworthy attitude summed up by another botanist who penned the lines:

There should be no monotony
In studying your botany;
It helps to train
And spur the brain—
Unless you haven't gotany.

If only all botanical authors were so considerate to their readers. However, there is a small list of just two obstacles I must bore you with. In a terrestrial environment, how can sperm swim and how can fertilized eggs avoid drying out?

Land plants tackle these problems in a variety of ways. Mosses and ferns, for example, cannot really be said to have solved the first problem at all, as they require damp conditions for sexual reproduction. In these groups sperm need a film of moisture to enable them to swim from male organs to female ones. This restricts the distribution of these plants to habitats that are at least occasionally wet. In ferns, the large, leafy plants that we are so familiar with are asexual and produce no eggs or sperm. Instead, they produce tiny, dust-like spores. When shed, these germinate to produce a microscopic sexual stage that leads an independent existence. This in turn produces the eggs and sperm. After fertilization, the resulting embryo takes root and develops into the large leafy plant we recognize as a fern. Some marine algae also have separate asexual and sexual stages, and this arrangement must also have been present in the ancestor of all land plants.

In the sixteenth century it was commonly and erroneously believed that ferns must reproduce by seed. But where were the seeds? Surely, since all plants grew from seed and the seeds of ferns could not be found, fern seed must be invisible! At that time herbalists believed that plants betrayed their medicinal uses in the shapes of their leaves and flowers, so kidney vetch was good for kidney complaints and liverwort benefited the liver. It was but a natural extension of the doctrine of signatures, as this herbal system was known, to assume that carrying invisible fern seed would confer invisibility upon the bearer.

Of course a problem stood between any herbalist peddling this theory and a huge money-making opportunity: how do you obtain fern seed? There was a way. Fern seed could be collected on the stroke of midnight on Midsummer Night's eve, but only by catching it as it fell from the plant onto a stack of twelve pewter plates. It would pass through the first eleven, but be trapped by the twelfth. Not everyone believed this, of course. In Shakespeare's play *King Henry IV*, written in 1597, a thief called Gadshill attempts to recruit an accomplice to a robbery by telling him that they will "steal as in a castle, cocksure; we have the receipt of fern-seed, we walk invisible." To which he receives the reply, "Nay, by my faith, I think you are more beholding to the night than to fern-seed for your walking invisible." No one nowadays believes in the existence of fern seed, but homeopaths still believe in the power of remedies containing herbal extracts that have been diluted to invisibility, so perhaps we shouldn't mock our credulous forebears.

Ferns and mosses may lack seeds and their reproductive cycles may be reminiscent of that of some marine algae, but there is one feature of their life cycle that they share with other land plants and which distinguishes all of them from algae: all land plants produce a multicellular embryo that is retained within maternal tissue. For this reason land plants as a group are known as embryophytes (i.e., plants with embryos). Precisely when during its development the embryo is released varies a great deal between species, but even the most negligent mother among terrestrial organisms never behaves the way many marine animals and plants do, which is to squirt their eggs and sperm into the environment and forget about them. Amphibians (frogs, toads, and salamanders) come close to this strategy, but of course they return to water to reproduce. It is no coincidence that land plants as a group can be defined as embryophytes. Parental care of the embryo is essential for successful reproduction on land.

The retention of fertilized eggs within maternal tissue where the developing embryo could be protected from desiccation was a crucial evolutionary step in the colonization of land from the sea, but free-living sexual stages like those of ferns still needed a wet environment to reproduce. Because we are looking back at the evolutionary history of the seed, it is almost impossible to resist describing what I am about to reveal of that history as "the next step" in seed evolution. But, though evolution seems to have a direction when we look back from the vantage point of the present, it doesn't follow purposeful steps—rather, it wanders from one chanced-upon solution to the next with no aim whatsoever. One particular path among the many wanderings of evolution led to the evolution of the seed. Others led to modern representatives of plants like mosses and ferns that have no seeds, and yet others led to extinction for plants like giant club mosses and seed ferns.

This caution issued, what happened next on the particular path that we are following to the evolution of the seed liberated plants from a dependence on a watery environment for sex. The large, robust plant, instead of shedding its female spores, retained them within its tissue where they became a protected, tiny sex machine.

Of course there were consequences of this for male spores. Once the female had become cloistered, males had to find their way to the egg by a different route. Swimming would no longer do it. Male spores were already equipped for aerial dispersal, though, so all that had to happen was for these to delay the liberation of their sperm until they reached the vicinity of the egg. Thus the male spore became a pollen grain.

The earliest seed plants found in the fossil record appeared in the Devonian period about 360 million years ago. They belonged to the group known as gymnosperms, whose living members include ginkgos, cycads, and conifers. The name "gymnosperm" is derived from the Greek for "naked seed," because seeds of these plants are not enclosed in an ovary. Incidentally, the word "gymnast" has the same root: classical Greek gymnasts performed naked. The maidenhair tree, *Ginkgo biloba*, is a living fossil gymnosperm whose reproductive system still contains more than a flavor of the sea recipe from which embryophytes evolved.

Ginkgo biloba is the last remaining representative of an ancient and once more numerous group of gymnosperms. Ginkgo ancestors are found in Permian fossil deposits 280 million years old. *G. biloba* was first found by a Western botanist in monastery grounds in China, but it now grows in botanical gardens and parks all over the world. This living fossil is a great survivor. One specimen survived the blast of the atomic bomb dropped on Hiroshima in 1945, even though it was only 1.1 kilometers from the epicenter of the explosion. The species is also extremely pollution tolerant. Many streets in New York City are planted with ginkgos, but only with males, because females produce mature seeds that have an unpleasant smell like rancid butter. No doubt the smell pleased the dinosaurs that once fed upon its seeds, but it is a deterrent to humans who have replaced them as dispersal agents. If you can find a mature female tree in spring you will see that its naked, unfertilized seeds dangle suggestively in pairs on the end of long stalks.

Male ginkgo trees produce wind-dispersed pollen grains that each contain an undeveloped male. When an unfertilized seed (called an ovule) on a female tree is ready for pollination, a drop of mucilage is exuded from a small pore in its tip. The mucilage is later retracted so that any wind-borne ginkgo pollen that has become trapped in the mucilage is drawn into a chamber inside the ovule. Males gather inside the chamber, still immature, each in his tiny flying saucer of a pollen grain. A marriage is now predestined between one of the waiting juvenile males and the equally immature female, but first they must mature and then there will be a contest between sperm. Had he known about it when he wrote his book-length poem about plant sex, *The Loves of the Plants*, Erasmus Darwin would undoubtedly have risen to the very heights of poesy in honor of this betrothal between juveniles with its prenuptial contest.

The arrival of pollen triggers the female cells within the ovule to begin development, but it takes up to four months before her egg is ready for fertilization. Any ovules that remain unpollinated are cast off by the female tree, who cuts her losses by severing the connection between stalk and branch. Meanwhile, inside the pollinated ovules the males within their chamber have also been developing, each living parasitically by drawing nutriment from the ovule through a tube. (Evolution will put this feeding tube to a different use in other seed plants, as we shall shortly see). Once her egg is mature, the female breaches the wall of the pollination chamber and floods it with liquid, creating a droplet of ocean in the ovule. Now ready for action, the pollen grains each discharge two massive sperm. Each one is powered by thousands of beating hairs called cilia that are arranged in spiral fashion around the outside of the sperm cell, forcing it forward like a marine torpedo. The winner of the sperm race fertilizes the egg and fathers the seed.

The motorized sperm of *Ginkgo biloba* were discovered by a Japanese botanist in 1896, and not long afterward a fellow countryman found that cycads had a similar sexual system with even bigger sperm powered by tens of thousands of cilia. Other gymnosperms have modified the sea recipe for sex still further, draining it of almost all its marine content and refashioning it for aerial service. The pollen grains of pines, spruces, and other conifers fly with the aid of wings. Conifer ovules, though still technically naked, are shielded between the scales of cones that flex open when the ovules are ready to be pollinated and shut again after fertilization. There is no simulated ocean in the ovule, or motorized sperm, but the tube that pollen grains use to parasitize the female tissue is there. In conifers, and in flowering plants too, the tube now serves a dual purpose: it feeds the male, but it also grows into a tubular conduit for the sperm to reach the egg. This is one of the countless examples in evolutionary history when a device that evolved in the service of one need (parasitic feeding) was turned to a quite different use (sperm delivery).

The acme of seed protection is reached in the flowering plants whose seeds are contained within an ovary. When the seeds begin to develop, the ovary around them matures into a fruit. The scientific name for flowering plants is "angiosperm," meaning "seed in a vessel." Not only are angiosperm seeds better protected than gymnosperm seeds, but they are nourished differently too. The food in a gymnosperm's picnic basket is supplied by the female tissue, rather as a human mother nourishes her unborn baby. However, the angiosperm embryo's picnic is provided by different, strange and even sinister fare—a tissue known as the endosperm.

In 1898 the Russian botanist Sergei Nawaschin announced that he had discovered that ovules in flowers he was studying were fertilized twice. Angiosperm pollen grains contain two sperm, just like in ginkgo, but unlike in ginkgo, Nawaschin discovered that in angiosperms both sperm find a mate. One sperm fuses with the egg cell and fathers the embryo, while the other sperm fuses with another nucleus that is found floating in the ovule. The product of this fusion becomes the endosperm, which grows into a food store. In some species, such as peas, the food in the endosperm is absorbed by the embryo during its growth, while in others, such as the grasses, it is not used until seed germination, when it feeds the seedling. Most of the kernel in cereal grains like corn, wheat, and rice is endosperm, and thus about 60 percent of the world's food supply is made up of this tissue.

Endosperm has been described as unique because it has three ancestors, but it leaves no descendants. Two of those ancestors are represented by the pair of chromosome sets that come from the female, the third is the single set of chromosomes from the pollen grain. This arrangement is very odd because embryos themselves only have two sets of chromosomes, one from each parent. Why should endosperm, which is just a food storage tissue, have three? The answer may lie in the mysterious evolutionary origins of endosperm.

There are three possible starting points for the evolution of endosperm. It could have arisen *de novo* with the origin of the angiosperms as an entirely new kind of tissue. This is highly unlikely because, as we have seen, evolution works by using what is already to hand to fashion new solutions. Every novelty has some kind of antecedent. Two different antecedent scenarios suggest themselves. One is that endosperm began as maternal tissue with two sets of chromosomes, after which double fertilization evolved and provided the third set. This is possible, but raises the question why endosperm would have two sets of chromosomes to begin with, when the female cells that produce and nurture the egg normally have only one set.

The other scenario is that endosperm began as an embryo, with one chromosome set from the egg and one from the sperm. Then at some point the maternal contribution doubled. This scenario is the sinister one because it suggests that the endosperm is the parasitized sibling of the developing embryo. Could the innocent-looking picnic basket hide a fratricidal secret: are angiosperm seeds raised on a diet of their own brethren?

One way to try to answer this question is to study the seed development of living fossils like ginkgo that might give us a clue as to which of the two scenarios is likely to have prevailed in the gymnosperm ancestor from which the angiosperms evolved. Ginkgo itself, which has no endosperm, is too distantly related to be of help, but there was a flurry of excitement in 1995 when double fertilization was discovered in *Ephedra*, a desert gymnosperm thought to share a common ancestor with the angiosperms. In *Ephedra* two eggs are fertilized, producing two identical embryos, each with one set of chromosomes from mom and one set from dad. Only one of the two embryos develops; the other always aborts. "Guilty!" rang the verdict when this discovery was made. It looked like double fertilization in the common ancestor of *Ephedra* and the angiosperms must both have produced two embryos, one of which turned into endosperm in the angiosperms.

"Not so fast," said counsel for the defense. This is only circumstantial evidence. What if *Ephedra* doesn't really share a common ancestor with the angiosperms? As in the best forensic crime mysteries, the solution hinged on DNA evidence. In 1999 new DNA data showed that *Ephedra* was not so closely related to the angiosperms after all. Case dismissed! *Ephedra* was a red herring. At the time of writing we are without a living witness to the momentous events that gave birth to the flowering plants. A living fossil gymnosperm that is closely related to the angiosperms may be out there somewhere, hidden like the Wollemi pine that was discovered as recently as 1994 growing in a vertiginous canyon in Australia, but it hasn't turned up yet and maybe never will.

Although we can't be sure, it is possible that endosperm is a sacrificial, sterilized embryo turned food-supplier to its sibling. One reason for thinking this is that a division of labor between carers for offspring (like the endosperm) and bearers of offspring (as the embryo is destined to become) is a common situation in nature. There are plenty of cases among social insects. Among honeybees, for example, the workers collect all the food for the hive and tend the queen's brood, but do not themselves reproduce. The queen produces eggs, but does not care directly for her offspring. At first sight this arrangement may seem to contradict Darwinian evolutionary principles. How can a sterile caste evolve if natural selection favors those individuals that leave the most descendants? Surely, sterile workers have no descendants, by definition?

Charles Darwin himself saw the problem, which he regarded as one of the most severe for his whole theory of evolution by natural selection. But he also saw the solution. "This difficulty, though appearing insuperable, is lessened, or, as I believe, disappears, when it is remembered that selection may be applied to the family, as well as to the individual, and may thus gain the desired end. Thus, a well-flavored vegetable is cooked, and the individual is destroyed; but the horticulturist sows seeds of the same stock, and confidently expects to get nearly the same variety." In other words, a sacrifice may sometimes pay off through the transmission of genes via relatives.

The closeness of the relationship between carer and bearer affects the likelihood that evolution will favor self-sacrifice by a carer. Human siblings share on average half their genes with each other, while first cousins who have two grandparents in common share one-eighth of their genes. Thus J. B. S. Haldane, one of the founders of modern evolutionary theory, quipped that he would lay down his life for two brothers or eight cousins. Honeybees belong to the hymenoptera, a group of insects with a peculiar sexual system that skews the relatedness among individuals within a hive. Male bees hatch from unfertilized eggs, so they have no father and carry only a single set of chromosomes, derived from mom. All a male bee's sperm are therefore identical, and when he mates with a queen bee, all of their female offspring get an identical set of genes from him. The result is that sisters in a beehive share not just half their genes (as in human families), but three-quarters of them. The queen honeybee mates just once in her life, so the workers in the hive are "supersisters" of the brood of larvae that they care for, sharing 75 percent of their genes with each one of them. This close relationship means that they can favor the transmission of their own genes to future generations by caring for their sisters. In Haldane's terms, a worker bee should lay down her life for one-and-a-third sisters.

The genetic peculiarities of honeybee families, about which Charles Darwin knew nothing, fully bear out his intuition that the evolution of worker bee sterility would be explained by benefits to other family members. To go further into the fascinating details of social insect evolution is beyond the purview of a book about seeds, but suffice it to say that explaining how social insects evolved, starting from an unpromising position as one of Darwin's chief difficulties, has become one of the crowning achievements of his theory.

The genetic logic that explained the evolution of sterile worker bees can be applied with equal success to the evolution of endosperm. The evolutionary stage has been set by worker bees; a different cast of characters is ready in the wings. The time is the dawn of the angiosperm era, and the curtain rises upon a new scene. Where there was a beehive, there is now an ovule and in it are two identical eggs. An insect arrives at the flower. She has only a walk-on part in this botanical drama, but she is bearing the principal male lead of the piece: a pollen grain. The pollen grain germinates, its pollen tube penetrates into the ovule, and two identical sperm slip down the tube into the ovary. Sperm meets egg, egg meets sperm and we have identical twin embryos. End of scene 1.

Scene 2: Two embryos in an ovule. The shade of J. B. S. Haldane strides onto the stage and utters the fateful line "If I were an embryo, I would lay down my life for an identical twin." A deathly hush pervades the swamp. Which embryo will become bearer, which will become carer, nobody knows, but whichever embryo sacrifices itself to feed the other, it will inherit the future in the genes of its well-fed twin. Curtain.

Thus, we can suppose that double fertilization produced two embryos in an angiosperm seed and a selfless sibling became an early endosperm, sacrificing itself to the appetite of its embryo twin. However, we know that the story did not end there, because at some point in angiosperm evolution the caring, sharing endosperm acquired a second set of maternal chromosomes. This changed the usual 1m:1p embryonic ratio of maternal (m) to paternal (p) genes to 2m:1p in the endosperm. How could this happen? Once more, the genetics of self-interest hold the answer.

Imagine a scenario soon after the evolution of endosperm. A pair of seeds is developing in a fruit. Seed 1 is swelling, drawing as much as it can upon maternal resources. Seed 2 is doing likewise. Should they share or should they compete? And who decides what proportion of limited maternal resources each seed should get? Let's try to answer this question from two different points of view: from the mother's perspective and from the father's. From the maternal perspective, each seed carries half her genes and there is no reason to prefer one of her offspring over another. So long as each seed receives enough resources to be viable, it is usually better to produce a number of seeds than to concentrate all available resources in one humongous seed. Evolution discovered that literally putting all your eggs in one basket was an unwise strategy long before humans coined this as a metaphor for risk. So, from the maternal perspective, resources should be shared equally among seeds.

Things look rather different from a father's point of view. All seeds on a plant have the same mother, but they do not all have the same father. Therefore, from a paternal perspective allowing resources to be shared among seeds will not increase an individual father's contribution to future generations. In fact quite the reverse, because resources relinquished for use by other seeds will only fatten the competition when seeds germinate and seedlings struggle with each other for light and nutrients. The difference in relatedness between parents and seeds therefore produces a conflict of interest between mother and father. It is in a father's interest to use his genes to grab all the resources he can for his seed. But it is in a mother's interest to share her limited resources among different seeds. How is this conflict resolved?

Endosperm plays an important role in mediating the conflict between parents over resource allocation to seeds because it is the supply line as well as the food store through which the mother nourishes the embryo. In a 1m:1p endosperm the sexes are evenly matched in the contest: one chromosome set each. In 2m:1p endosperm, the mother trumps the father and wrests genetic control from him by doubling her arsenal of genes. The benefit this brings in terms of the mother's seed output and her contribution to future generations is presumably why 2m:1p endosperm was favored by natural selection and why it is almost universal among flowering plants.

If you are not used to the idea yet that there is parental conflict in every kernel of an innocent-looking bag of popcorn, then you will quite rightly want some evidence. The best evidence is that when the relative doses of maternal: paternal genes are manipulated, this affects how seeds are resourced. Genetic experiments with corn have been used to manipulate the usual 2m:1p ratio of maternal to paternal contributions to endosperm to obtain different ratios. If evolution favored the change from an original 1m:1p to the now usual 2m:1p because the double dose of maternal genes controls seed size, then changing the ratio even further in the maternal direction should produce smaller corn kernels than the norm. This is indeed what has been found. Endosperm with an extra maternal chromosome set giving a ratio of 3m:1p produced smaller seeds than normal. A 4m:2p ratio restores the ratio to an effective 2m:1p, and these kernels were normal in size. This indicates that it is the balance between maternal and paternal chromosome sets and not their actual number that matters.

The idea that maternal and paternal parents have different effects on endosperm provisioning is further supported by experiments that have shown that there is a gene on one particular chromosome that is required for the production of normal endosperm in corn. The gene is present in both eggs and sperm, but only if it is inherited from dad does it work and produce normal-sized endosperm. The copy of the same gene present in the egg appears to be switched off. This is more evidence of genetic conflict between male and female parents over how resources are to be shared among seeds. So sex can bring conjugal conflict as well as connubial bliss, but there's no getting away from it. Sex is everywhere. Even beans do it.

{ 3 }

Even Beans Do It

SEX

Birds do it,
Bees do it,
Even educated fleas do it . . .
People say in Boston even beans do it . . ."
COLE PORTER

Seeds are the products of sex. The essence of sex is the exchange of genes between individuals. Everything else we associate with it, like males and females, seeds and sperm, pistils and penises, roses and romantic candlelit dinners are by comparison mere refinements—the frippery of evolutionary development—what evolution got up to when it had a lot of time on its hands. The big evolutionary breakthrough that involved two individuals exchanging DNA occurred very early in the history of life, before even the different roles of male and female had evolved. The very antiquity of the event is why virtually all life engages in sex. We all do it because all our ancestors did it, and of course we exist for that very same reason. Why, though, was sex so successful as a means of reproduction? Why has it persisted since almost the dawn of life? That is a puzzle.

The evolution of sex is a puzzle because, at least to the dispassionate observer, it seems such an inefficient way to transmit your genes to future generations. Why share your offspring with a mate, diluting your genetic legacy to each child by a half, when the alternative of reproducing asexually would mean that every child would be a Mini-Me? Sexual reproduction has been compared to a game of roulette in which the players throw away half their chips at every spin of the wheel. It works, but only if all players refrain from cheating. Yet, this seemingly handicapped sexual gamble pays off. What stops asexual cheats being successful and causes sexuals to win the game?

Nearly all animals as well as plants reproduce sexually, so the puzzle is not a particularly botanical one, but unlike most animals most plants do have asexual as well as sexual means of reproduction, so the persistence of sex in plants does seem especially odd. When strawberry plants are so good at spreading by runners, why should they produce seeds and fruit as well?

The realization that plants do reproduce sexually, and that this is what flowers are all about, is not as old as one might imagine. Greek and Roman philosophers were at best coy on the subject of plant sex, if not plain ignorant. The Greek philosopher Theophrastus wrote of male and female date palms, but if he believed that plants in general were sexual, he did not allude to this in his eighteen volumes of botanical writing. The Roman poet Ovid published a sex manual in verse, *The Art of Love (Ars amatoria)* that had him banished by the Emperor Augustus in a cleanup of Rome. In an attempt to rehabilitate himself and to show how religiously upright he was, Ovid wrote the work for which he is now most famous, the *Metamorphoses*. This retells legends of how the gods punished erring deities and mortals by transforming them into other beings. When a god's love went unrequited, the object of his or her lust was often punished by being transformed into a supposedly sexless plant: "If I cannot have you, none shall!" was the very human motivation for such ungodly revenge.

Narcissus suffered this fate for being so enamored with himself that he rejected the advances of the nymph Echo. Presumably unknown to Ovid or Echo, but as testified by their showy flowers, daffodils (*Narcissus* species), have a fascinatingly varied sex life that would not have been out of place in Ovid's own *Ars amatoria*. Thanks to godless science we now know that *Narcissus* had the last laugh after all. Hyacinthus was the unlucky casualty of rivalry for his affections between the gods Apollo and Zephyrus. When Hyacinthus died in the tangled knot of this love triangle, the flowers which bear his name sprang from the ground where his blood fell. Hyacinthus may have died a virgin, but his botanical reincarnation produces seeds aplenty.

Apollo, who evidently swung both ways, also took a liking to the beauty Daphne. She took fright and fled "as the lamb flees the wolf, or the deer the lion, or as doves fly from an eagle." The god, on fire with love, pursued her, and when Daphne felt his hot breath upon the back of her neck she cried out to her father, Peneus the river god, to destroy her beauty and thereby save her from Apollo's lust. Her father obliged and Daphne became an inviolable tree. Do think of her when you encounter the wonderful fragrance of the shrub *Daphne odora* and see insects pollinating its delicate pink flowers.

Ovid's *Ars amatoria* gives every kind of advice to young lovers, including where in Rome the best pickup spots are, but he says don't waste your time on the herbal aphrodisiacs recommended by some:

Nor drugs nor herbs will what you fancy prove,
And I pronounce them pois'nous all in love.
Some pepper bruis'd, with seeds of nettles join,

> *And clary steep in bowls of mellow wine:*
> Venus *is most adverse to forc'd delights*
> *Extorted flames pollute her genial rites.*

And, Ovid concludes, "beauty and youth need no provocative." Plants were neither sexual nor aphrodisiac.

Not till the dawn of the Enlightenment in seventeenth-century Europe did sexual reproduction of plants or animals become the subject of scientific enquiry. The starting pistol was fired in 1651 by William Harvey, discoverer of the circulation of the blood, who set down in his *Treatise on Generation* the principle that all life comes from eggs (*Omne vivum ex ovo*). In 1676, the physician and botanist Nehemiah Grew gave a lecture on the anatomy of flowers to the Royal Society of London in which he identified the stamens as the source of plant sperm. In further observations of plant anatomy that he published in 1683 Grew referred to "the foetus or true seed."

Experimental proof of the sexual functions of flower parts followed not long afterward when in 1694 the German botanist Rudolf Jakob Camerarius published his *Epistola de sexu Plantarum* (Letter on plant sex). Among various experimental results Camerarius reported that removing the female silks of corn plants resulted in no seed being set. James Logan, chief justice and president of the Council of Pennsylvania, probably conducted the first quantitative experiment of this kind, which he described in a letter sent in 1735 from Philadelphia to a fellow of the Royal Society in London. It was published under the title "Some Experiments concerning the Impregnation of Seeds of Plants" and reported that when Logan removed a half, a quarter or an eighth of the silks from an ear of corn he obtained cobs with exactly the corresponding proportion of a half, a quarter, or an eighth of the seeds missing.

When Carl Linnaeus, later to be dubbed the "Prince of Botany," became a student at Uppsala University in Sweden at the end of 1728, the sexuality of flowers was still not widely recognized. Linnaeus studied what had been written on the subject and gathered his thoughts in an essay that he submitted to his professor as a New Year greeting for 1730. It was customary for students' New Year greetings to be written in verse, but Linnaeus wrote: "I am not born a poet but somewhat of a botanist and because of this, present a fruit of the year's small crop that God has bestowed on me. . . . In these few pages is treated the great analogy which is to be found between plants and animals in that they increase their family in the same way," and he continued:

The petal of the flower in itself contributes nothing to generation but only serves as the bridal bed, which the Great Creator arranged so beautifully, and garnished with such precious bed-curtains, and perfumed with so many delicious scents, in order that the bridegroom with his bride may therein celebrate their nuptials with so much greater solemnity. When the bed has been so prepared, is the time for the bridegroom to embrace his darling bride, and loose himself in her. . . . Watch with me how one flower breaks out of its calyx and again another from a bud! Watch further how the one sort of plant (with regard to build, form and look) can in a thousand ways exhibit likenesses and differences! . . . It is not said by poets without reason, that from the blood of the eternal gods many plants have sprung up.

The "likenesses and differences" between the sex organs of flowers were to become the organizing principle of Linnaeus's classification of plants, which he included in his book *Systema natura*, first published in 1735. So successful was *Systema natura*, which covered animals as well as plants, that it grew from a mere fourteen sheets in the first edition to three volumes totaling 2,300 pages by the time the twelfth edition was published little more than thirty years later. In England, Erasmus Darwin had great success with his poetic rendering of Linnaeus's taxonomy, called *The Loves of the Plants*, and first published in 1789.

Today, Linnaeus is remembered for turning the classification of living things into a rigorous science, for the binomial names that are used for the scientific designation of species (e.g., *Daphne odora*, *Narcissus pseudonarcissus*) and for the innumerable species he named, rather than for his sexual classification of plants. Botanists found that the Linnaean sexual system of classification failed to fit the natural relationships among plants that became evident even before the evolutionary source of those relationships was recognized. Ironically, the Prince of Botanists was more successful with animals in this respect, correctly recognizing that whales and bats are mammals and that humans are primates.

Though the fact of plant sexuality had been established by the middle of the eighteenth century, there was still debate about the respective roles of egg and sperm in the production of the embryo. For a while there were two contending camps on the issue. On the one hand were the ovists, who believed that sperm merely stimulated the egg to bring forth its embryo without making any substantial hereditary contribution to it. In 1727 Henry Baker summed up this view admirably in verse:

Each Seed includes a plant, that Plant, again,
Has other Seeds, which other Plants contain:
Those other Plants have All their Seeds, and Those
More Plants again, successively, inclose.
Thus ev'ry single Berry that we find,
Has, really, in itself whole Forests of its kind . . .

The opposite view was held by the spermists, notable amongst them the Dutchman Antoni van Leeuwenhoek, who was the first to observe sperm under the microscope. Spermists believed that the egg was a mere receptacle and carer for the embryo, which they thought was actually transmitted by the sperm. This idea can be traced back to the ancient Greeks. In *The Eumenides* by Aeschylus, the god Apollo defends Orestes from the charge of matricide by stating as an irrefutable fact that "the mother is not parent of her so-called child but only nurse of the new-sown seed. The man who puts it there is parent; she merely cultivates the shoot."

The Dutch spermist Nicolas Hartsoeker believed he could actually discern a tiny human form folded in the head of a sperm, and he published a drawing of this homunculus in 1694. Spermism had a brief life (just as sperm do), but its botanical equivalent, pollenism, endured a while longer. One proponent was Sir John Hill, an extraordinary character of a kind that probably only eighteenth-century London could have produced: herbalist, physician, prolific journalist, scientific translator, playwright, astronomer, geologist, microscopist, botanist, and actor. He was even thought to be the pseudonymous author of a recipe book, *Mrs. Glasse's Cookery*. Hill's enormous industry was undermined by a seeming determination to quarrel with anyone and everyone of note, especially his erstwhile friends. One friend he fell out with was the actor and theater manager David Garrick, who wrote of his doctoring and playwriting:

Thou essence of dock, valerian and sage,
At once the disgrace and the pest of the age;
The worst we can wish you for all your dam'd crimes
Is to take thy own physics and read thy own rhymes.

Among the seventy-six publications Hill produced in his lifetime was a twenty-six volume work containing 1,600 copperplate engravings called *The Vegetable System*, in which he expressed the pollenist view that the seed is a receptacle for the embryo that is delivered to it by the pollen grain. In this he was, of course, just as wrong as were the ovists. A man of half Hill's accomplishments who was not so full of himself and so ready to offend would easily have been elected a fellow of the Royal Society, but he was distrusted even when he was right. He took rejection by the Royal

Society in typical fashion, publishing satires that ridiculed the fellows. One was called "*Lucine sine concubitu* [Pregnancy without intercourse], a letter humbly addressed to the Royal Society; in which it is proved, by most Incontestable Evidence, drawn from Reason and Practise, that a Woman may conceive, and be brought to Bed, without any Commerce with Man."

In the 250 or so years since John Hill spurned and was spurned by the Royal Society, science has revealed marvels in every sphere, but scientists have become pale and colorless by comparison, at least in print. I did once read a private letter from a reviewer to the editor of a journal which suggested that among the nine authors of a paper that was based on only three days of field observations, "half the authors remained in some rural pub, for the duration, downing pints and ogling the norbs on the nubile daughters of local wheat farmers." But that's the most fun I've had in twenty years of reading review letters. I can't even reveal the name of the person who wrote it, because if he ever found himself in the wrong pub it would be nine against one and I couldn't have that on my conscience.

If the dubious satisfaction of eighteenth-century knockabout satire is denied us, then we'll have to make do with twenty-first-century irony. It is surely ironic that we now know that there are indeed plants that get pregnant (so to speak) without any "commerce" with a male. It is doubly ironic that there are also a few known cases of what appears to be genuine spermism, although without the homunculus. So is it one-all in the battle, Royal versus John Hill? Not quite.

To be fair, there is only one known case of spermism in plants, or to give it its correct scientific name, *androgenesis*—literally, "birth from a male." This was discovered quite recently in the very rare Sahara cypress *Cupressus dupreziana*, which is one of the most endangered plants in the world. The entire species is reduced to only about 230 trees scattered among a handful of desert oases in the Algerian Sahara. The first sign that something odd might be happening in Sahara cypress was the finding that the sperm cell in its pollen grains carried a double set of chromosomes, not the single set that is normal for sperm. Genetic fingerprinting of seeds then revealed that the embryo inside them appeared to be unrelated to the tree on which they were produced. It turns out that the embryo in the seeds of the Sahara cypress is a clone of its father, with no genetic relationship to the mother tree which bears it. This is exactly the situation described by Aeschylus: "The mother is not parent of her so-called child but only nurse of the new-sown seed. The man who puts it there is parent; she merely cultivates the shoot." Indeed, the reproductive habits of the Sahara cypress even have consequences that would fit a Greek tragedy.

When the pollen of *Cupressus dupreziana* was then used to fertilize another species of *Cupressus*, the offspring were identical to *C. dupreziana* and not hybrids between the two species as would normally be expected. This is the first reported case of surrogate motherhood in the plant kingdom. Somehow, the sperm of *Cupressus dupreziana* is able to supplant the maternal chromosomes in the egg cell of the ovule, so the embryo becomes a clone of its father. This is genetic parasitism, more akin to piracy than the charitable act that the term "surrogate motherhood" might imply.

It is interesting to speculate whether the extreme rarity of Sahara cypress might be linked to its peculiar reproductive habits and what this might mean for the future of the species. Only 10 percent of seeds of the Sahara cypress contain a viable embryo. Although some pollen grains contain sperm with twice the normal complement of chromosomes and these can make embryos by androgenesis, other pollen grains are even more abnormal and are therefore sexually sterile. If there are no normal sperm being produced, then no normal seeds can be set either.

Could the Sahara cypress survive entirely through androgenetic reproduction? If there were surrogate mothers of other cypress species available in its native habitat, perhaps the Sahara cypress could become a cuckoo of the desert, surviving by using other species to raise its offspring, but sadly there are none. How long could Sahara cypress survive in isolation? Cypresses, like other conifers, produce both male and female cones on the same tree, but these are costly to produce and there is likely to be a trade-off between the number of cones of each sex that are made. If a tree produced fewer female cones, it could produce more male ones.

Because androgenesis is this tree's only means of reproduction, over the generations evolution will strongly favor trees that produce more male cones and fewer female ones. As female cones dwindle in frequency the number of seeds being produced will also fall, pushing the population nearer and nearer the brink of extinction. Indeed, this may be how the population found itself on that brink already, although humans have certainly helped it to the precipice by chopping down trees for wood.

If we could resurrect Sir John Hill and tell him about androgenesis, he could scarcely claim to be vindicated, but we could rebury him with honor in a coffin of nearly extinct cypress wood. Perhaps the irony would amuse his spirit, for he who trod the boards of the Garrick Theatre would be familiar with the Clown's song from William Shakespeare's *Twelfth Night*:

Come away, come away, death,
And in sad cypress let me be laid;
Fly away, fly away, breath,
I am slain by a cruel maid.

The sad Sahara cypress is nearly slain, not by a cruel maid, but by a selfish gene. You may wonder how evolution, which favors survivors in the Darwinian struggle for existence, could push a species to the brink of extinction. How could evolution endanger a species? The answer illustrates a fundamental principal of evolution: natural selection works for the good of the individual, not for the good of the species. Usually, anything that increases the number of descendants that an individual will leave will also increase, or at least maintain, the total number of individuals in a population, and so we are unused to the counterintuitive idea that evolution could threaten species survival by bringing about a lowering of total population size. However, in the unusual predicament that the Sahara cypress finds itself in, this is quite possible, perhaps even inevitable.

The problem is caused by the fact that in androgenesis all descendants arise from pollen; although ovules are required for the formation of seeds, they do not propagate the genes of the tree that produces them. Even though conifers, like the majority of plants, are hermaphrodites and have both male and female organs, they do not self-pollinate. Thus, when androgenesis takes hold in a population, no tree can increase the number of offspring carrying its genes by producing female cones. Only male cones producing pollen can do this, so any gene that favors male cone production will spread. Because of the trade-off in production of male and female cones, the result of increasing the number of male cones will be a reduction in female cone production. Fewer female cones means fewer seeds and this means fewer future trees.

One of the reasons this situation is unfamiliar to us is that as social animals we have a way out of the kind of predicament that brings the threat of extinction down on the head of selfish cypresses. Social animals can gain rewards for seemingly altruistic acts, such as raising another's children, by receiving reciprocal acts of kindness. It is the strategy of "You scratch my back and I'll scratch yours." Lacking any social system, Sahara cypress cannot say "I'll produce female cones to raise your pollen in if you'll do the same for me," even though such a strategy would be consistent with natural selection and could save the Sahara cypress. Some hermaphroditic fish do this.

There is no evolutionary future in being a male if there are no females, but all-female populations are perfectly viable. Indeed, plants that produce seeds without fertilization are quite common. The common dandelion (*Taraxacum officinale*) has populations that produce seeds asexually, and this is also known in blackberries (*Rubus* spp.), whitebeams (*Sorbus* spp.), hawthorns (*Crataegus* spp.), and many other groups. Asexual production of seed is called "apomixis," and its products have a double set of maternal chromosomes instead of the usual one set from a male and one from a female parent. Just to confuse botanists, seed set in some apomicts still requires pollination to stimulate seed development, but sperm never actually contribute to the genetic inheritance of the seed.

All the seeds produced by an apomictic plant are genetically identical to their mother, from which you might expect entire populations of dandelions to be genetically uniform, but this is rarely the case. It is as if apomictic dandelions every now and again stumble into sexuality, generating new lines of clones that then make copies identical to themselves. Occasionally almost a whole species, like the blackberry *Rubus nessensis* native to Northern Europe, appears to belong to a single apomictic clone, but this is very rare. For some reason, even apomictic plants have trouble giving up sex altogether.

Apomixis demonstrates that evolution repeatedly attempts to break the habit of sex in plants, but these attempts are never wholly successful. It's easy to see why androgenesis is a no-hope strategy if you are going to reproduce without sex, but it's a mystery why apomixis isn't the norm in plants once the dependence upon pollination has been broken. Some clues as to what might go wrong for asexually reproducing plants can be gleaned by looking at the genetic makeup of plant populations that, like the strawberry, can reproduce clonally and sexually. Because such plants have both options open to them, you can ask, are there particular environments or certain circumstances where sex is favored over clonality?

But first how can you tell whether a population was born

mainly out of sexual reproduction or from clones? The genetic fingerprint of an individual, called its genotype, is the key. Sexually produced offspring generally have different genotypes from one another, while clones are genetically identical. So if you sample the genotypes of one hundred individuals in a population and all are different, then each must have started life from a different, sexually produced seed. On the other hand, if the one hundred plants sampled all have the same genotype, then that particular population must have arisen from asexual (clonal) reproduction. Intermediate numbers of genotypes indicate a mixture of sexual and asexual reproduction. What are the patterns found in nature?

I surveyed the results of several hundred studies of the genetic composition of plants that are capable of both sexual and asexual reproduction and this showed some surprisingly clear patterns. First, populations that consist of just one genotype, indicating that sexual reproduction has played little or no role in their growth, are very unusual among terrestrial plants, except in species that are rare or endangered. Small population size, which affects many rare and endangered plants, leads to inbreeding and this often adversely affects seed production. Therefore, perhaps the reason these species are rare in the first place is because they are sexual failures and they are just clinging on to existence thanks to their ability to reproduce by suckers or other vegetative means. In other words, dependence on asexual reproduction in populations of endangered plants is a sign of sexual failure rather than of asexual success.

The situation is different in aquatic plants, among which exclusively asexual populations are comparatively common. The reason for this is probably that vegetative fragments can be carried long distances in water and these can become the founders of new populations that then multiply asexually. Vegetative fragments are not so easily dispersed on land, and here it is seeds, equipped with wings, parachutes, or animal help, that are better dispersed. Since seeds are usually sexual products, their superior power of dispersal means that even new terrestrial populations contain many genotypes.

The other plants that depend more than the average on asexual reproduction are non-native aliens. Here, the ability to reproduce without a mate may make asexual reproduction especially useful to invading species. Japanese knotweed *Reynoutria japonica* is a good example. All the alien populations of this species in Britain, continental Europe, and the United States consist of the same single asexual clone!

Apomictic seeds, like those of dandelions or blackberries, have the same aids to dispersal as the sexual counterparts from which they evolved, so it is not surprising that they have no trouble getting around and that asexual populations of these species contain comparatively few genotypes. But, as already mentioned, apomicts do occasionally have sex, and so even dandelion populations usually contain more than a single genotype. Apomicts are peculiarly common at the edge of species' ranges. Again, this is possibly a sign of sexual failure in unfavorable conditions.

Collectively, these patterns suggest that the ultimately successful asexual plant would be a rare, alien, aquatic apomict living in a geographically marginal habitat. There is no single species that fits this Identi-Kit image, if only because successful aliens are by definition not rare, but the picture painted does suggest that asexual reproduction takes off only in certain very specific ecological circumstances. It reinforces the message that it is sex that is the real universal success. Why?

Dozens of theories have been dreamt up to explain how sex overcomes its apparent handicap in the roulette game of life. Most of these have foundered on the problem that they are not general enough to explain why sex is universal. These problems have even caused some scientists to despair that any general explanation for the evolution of sex exists, leading to the conclusion that there must be lots of different answers to this one big question. At the time of writing, there is still no consensus about a solution, but the evidence does now seem to be pointing toward two particular theories. Ironically, one of them is essentially the same as the very first idea that was ever offered on the subject by Thomas Hunt Morgan, later to become a Nobel laureate, in a lecture on the evolution of sex at Columbia University in New York in 1913.

At the time of Morgan's lecture the term "gene" was not yet in common currency, but in modern terms what Morgan suggested was that the advantage of sexual over asexual reproduction is that sexually produced offspring can inherit an accumulation of beneficial genes from a wide network of ancestors, while all asexuals get is what their mother had. Copies of beneficial genetic mutations are multiplied by natural selection and accumulate over the generations. Through sexual reproduction they can become combined with other beneficial mutations and increasingly concentrated in the offspring. A sexually produced offspring has two parents, four grandparents, eight great-grandparents, sixteen great-great-grandparents, and so on into the remote past. This ever-widening network of ancestors is like an immensely deep funnel through which any favorable genes that arose in the past are collected together and passed down to the most recent generation. The links in the network of ancestors are forged by sex.

If there is no sex, there is no cornucopia of beneficial genes funneling into each new generation. Asexual offspring have no network of ancestors, but just a lineage, like a single-file line of identical clones stretching into the past. They have one parent, one grandparent, one great-grandparent, and so on back to some remote, single ancestor that first slipped from the ancient habit of sex. Asexuals can succeed in the short term, and indeed, like alien populations of Japanese knotweed, they can spread with stupendous success, but they lack the raw material of evolution: genetic variation. Without genetic variation evolutionary change is brought to a standstill and sooner or later conditions will arise that require an adaptation that is simply not to be found in the asexual population.

The greatest danger for asexual populations is possibly disease. In the 1970s elm trees right across Europe and North America were destroyed by a fungal disease transmitted by elm bark beetle. The epidemic was caused by a new strain of the fungus that was probably spread with imported logs. Some elms were much more badly affected than others, indicating that genetic variation among species played a role in susceptibility. In Britain the worst affected species was the English elm, *Ulmus procera*. This species had long been a mystery because it never set seed and spread only by underground suckers. Recent genetic analysis has discovered that the entire species was in fact a single clone that the Romans brought to Britain two thousand years ago to provide support for vines in their vineyards and which they propagated by cuttings. Because *Ulmus procera* was entirely clonal, there was no genetic variation for resistance to elm disease and all trees were virtually wiped out by it.

The second theory that also now looks like a promisingly universal solution to the problem of sex is complementary to the first. According to Morgan's theory, asexuals cannot benefit from the inherited accumulation of beneficial mutations because they are genetically isolated from all lineages but their own. This means they cannot evolve and adapt. According to the second theory, suggested by H. J. Muller in 1964, the same genetic isolation also means that deleterious mutations can accumulate within a clonal lineage. In a sexual population natural selection can weed out individuals carrying an especially heavy load of mutations, leaving undamaged offspring to continue the line. But in asexual populations individuals with accumulated mutations exist in multiple copies, and selection cannot purge deleterious mutations just by removing the few most severely affected offspring: whole clonal lineages must be removed.

In fact, because mortality is not always driven by natural selection, but can also occur through chance events, clonal lineages with just a few deleterious mutations will also occasionally die out too, causing the average number of mutations in the remaining population to rise. The process by which an ever-increasing load of mutations accumulates in asexual populations is called "Muller's ratchet" because, like the tool that will only turn in one direction to become tighter and tighter, the mutation load can move in only one direction—up.

Thus, once a population starts to reproduce asexually, it may have trespassed onto a slippery slope, and accumulating deleterious mutations may cause it to lose the ability to reproduce sexually altogether. This may be what has happened in populations of the North American aquatic, swamp loosestrife *Decodon verticillatus*, which is sexually sterile in several populations at the margins of its geographical range. In this particular case natural selection appears to have helped the species over the edge into asexuality because vegetative growth is more vigorous in sterile plants than sexually fertile ones. This trade-off between sexual and asexual reproduction would favor a shift to asexuality in a population where few seeds could be set because it was at the edge of the climatic zone to which it was adapted.

Despite the promising signs that the evolutionary problem of sex may soon be solved, there seems no danger that it will lose its fascination or mystery, particularly for anyone interested in plants. There is more sexual variety in how plants make seeds than there are pages in *The Joy of Sex* or positions in the *Kama Sutra*. As the poet E. B. White put it, "Before the seed there comes the thought of bloom," and it is to some fascinating flowers that we now turn our thoughts.

{ 4 }

Before the Seed

POLLINATION

Before the seed there comes the thought of bloom.

E. B. WHITE

According to his amanuensis Dr. Watson, Sherlock Holmes, the man who brought the utmost scientific rigor to criminal detection, had almost supernatural powers of observation. The slightest telltale details of a person's appearance allowed the fictional detective to reconstruct the person's profession, deduce his or her recent travels, and even sometimes to penetrate a secret identity. How strange then, that in the mystery *The Naval Treaty*, Holmes interrupts his sleuthing with this reverie prompted by the sight of a rose outside the window: "Our highest assurance of the goodness of Providence seems to me to rest in the flowers. All other things, our powers, our desires, our food, are really necessary for our existence in the first instance. But this rose is an extra. Its smell and its colour are an embellishment of life, not a condition of it. It is only goodness which gives extras, and so I say again that we have much to hope from the flowers."

Dr. Watson reports that the great detective's clients, patient witnesses of the performance and clearly hoping for more practical assistance from him, "looked at Holmes during this demonstration with surprise and a good deal of disappointment written upon their faces. He had fallen into a reverie, with the moss rose between his fingers. It had lasted some minutes before the young lady [client] broke in upon it. 'Do you see any prospect of solving this mystery, Mr. Holmes?' she asked, with a touch of asperity in her voice."

Needless to say, Holmes does crack the mystery in the end, but as for the case of the rose which he regarded as a purposeless extra to life, he was on the wrong track entirely, and we can all share in the disappointment at his uncharacteristic lack of perspicacity. When *The Naval Treaty* was first published in the *Strand* magazine in 1893, Holmes's antiquated view of flowers had long been contradicted by evidence that their true function is as sexual ornaments that aid pollination.

Charles Darwin, a reasoner with powers more uncanny even than those of the fictional Sherlock Holmes, wrote in his book entitled *The Effects of Cross and Self-fertilization in the Vegetable Kingdom* published in 1876, "There is weighty and abundant evidence that the flowers of most kinds of plants are constructed so as to be occasionally or habitually cross-fertilized by pollen from another flower" and "flowers are adapted for the production of seed and the propagation of the species." Darwin was especially fascinated by orchids, whose highly conspicuous flowers are fashioned by natural selection into extreme shapes and extraordinary dimensions, all in the service of producing seeds that are so inconspicuous and tiny that they are almost as invisible as "fern seed."

Among Darwin's many feats of Sherlock Holmesian deduction was his prediction, more than forty years before its actual discovery, of what the pollinator of the Madagascan star orchid *Angraecum sesquipedale* would look like. The species name of this plant (*sesquipedale*) means in Latin "a foot and a half," which, with piscatorial exaggeration, describes the foot-long nectar tube that hangs like a whiptail from the back of the flower. In his book, *The Various Contrivances by Which Orchids Are Fertilized by Insects*, Darwin recounts:

In several flowers sent me by Mr. Bateman I found the nectaries eleven and a half inches long, with only the lower inch and a half filled with nectar. What can be the use, it may be asked, of a nectary of such disproportionate length? We shall, I think, see that the fertilization of the plant depends on this length, and on nectar being contained only within the lower and attenuated extremity. It is, however, surprising that any insect should be able to reach the nectar. . . . In Madagascar there must be moths with proboscides capable of extension to a length of between ten and eleven inches! This belief of mine has been ridiculed by some entomologists.

As Sherlock Holmes famously remarked to Dr. Watson in *The Sign of Four*, "How often have I said to you that when you have eliminated the impossible, whatever remains, *however improbable*, must be the truth?" A giant Madagascan hawk moth with an improbably long tongue was reported in 1903. It was a subspecies of an African moth *Xanthopan morganii* and was given the name *Xanthopan morganii praedicta*. Not until the late 1990s was the moth actually observed in action, behaving just as Darwin had predicted it would more than a century earlier.

Darwin's fascination with flowers was not a Victorian country gentleman's whimsical pursuit of natural history for its own sake, but part of a systematic and comprehensive research program into how natural selection could explain the details of adaptation. Charles Darwin's son Francis described how the program began when a botanist friend named Robert Brown sent his father a then little-known book in German by the Lutheran cleric Christian Konrad Sprengel called *The Secret of Nature Revealed in the Structure and Fertilization of Flowers*. Francis Darwin wrote that the book "not only encouraged him in kindred speculation, but guided him in his work. . . . It may be doubted that Robert Brown ever planted a more fruitful seed than in putting such a book into such hands."

Darwin had the leisure to pursue his studies that Sprengel had not. The cleric spent so much time on his research that he was dismissed from his post for neglecting his flock. Sprengel described insects visiting flowers, but he did not explain *why* it was important for the pollen from one flower to be transported to the stigma of another. Why didn't flowers, which more often than not contain both male and female organs, simply pollinate themselves? For want of an answer to this vital question, Sprengel's book failed to excite the interest that its title clearly suggests he thought it deserved. Darwin, however, found the answer and thereby revealed the true significance of Sprengel's observations. Sprengel was posthumously celebrated for the revelations in his book.

Though Darwin believed there must be an advantage to cross-fertilization in plants, he at first thought that the obvious experiment of comparing the progeny of self-fertilization with plants raised from cross-fertilization was unlikely to show any difference between them, since he knew it rarely did so in the first generation of crosses between animals. As a result, his first observations of this kind were made by accident in plants that he had bred for experiments on inheritance. In these first, unplanned experiments it turned out that "self-fertilized seedlings were plainly inferior in height and vigour to the crossed." This chance observation caused Darwin to begin a mammoth program of experiments in dozens of species comparing the growth of plants raised from self-fertilized seeds with the progeny of matings between different individuals. To this day no one has repeated experiments of this kind on the scale that Darwin accomplished. The experiments showed quite unequivocally why it is to the advantage of plants to produce elaborate flowers that facilitate outcrossing: self-fertilization produces inferior offspring.

The inferiority of inbred offspring is now recognized as a very general phenomenon known as inbreeding depression. The adverse affects of inbreeding had a personal as well as a scientific interest for Darwin. First-cousin marriages in the Darwin and Wedgwood families had become a pattern, linking the two families. Charles's mother was Susannah Wedgwood, daughter of Josiah Wedgwood I, who founded the famous pottery of the same name. Charles married Emma Wedgwood, his first cousin and daughter of Josiah Wedgwood II. Four of Josiah Wedgwood II's seven children who reached adulthood married first cousins, including Emma who married Charles and her brother Josiah III who married Charles's sister Caroline Darwin.

Acutely aware of the importance of heredity and the power of natural selection to weed out the weak and sickly, Darwin anxiously watched the health of his growing family and readily ascribed their ill health to inheritance of his own weak constitution. When his beloved eldest child Annie died at the age of ten it seemed to confirm his worst fears, and he confided in a letter to a relative that "my dread is hereditary ill-health." It is now thought that Annie probably died of tuberculosis and that Darwin's own ill health was caused by a parasitic disease he contracted on his travels in South America. Neither illness was hereditary.

Darwin was worried that consanguineous marriages like his own might be injurious to the health of the nation, and in typical fashion he sought data to test this idea. He asked the British parliament to consider adding a question to the national census of 1871 that would ask whether respondents were married to first cousins. Answers to this question could then be cross-referenced with the number of surviving children reported on the census form to determine whether first-cousin marriages were less fertile than those between unrelated partners. Parliament hotly debated the issue, but threw it out on a vote of two to one against on grounds of civil liberties. Among the objections voiced by members of parliament were that children would be "anatomised by science" as though they were plants or animals. Of course to Darwin, the same biological laws applied as much to humans as to animals or plants. A scientific enquiry that may have appeared callous to parliamentarians was no doubt motivated on Darwin's part by a desire to spare others the suffering he had experienced in his own family.

If progress could not be made with human subjects, the news was much better in the realm of plants, where evidence of adaptation for outcrossing was beginning to build up. In 1873 the state entomologist of Missouri, Charles Valentine Riley, reported in the *American Naturalist* the extraordinary way in which yucca plants were pollinated by moths that parasitize their seeds. Darwin read the report and wrote to his friend Joseph Hooker, the director of the Royal Botanic Gardens at Kew, that this was "the most wonderful case of fertilization ever published."

There are about thirty species of yuccas, all native to North America and ranging from the South of Mexico to the U.S.-Canadian border. *Yucca filamentosa* and *Y. gloriosa* are widely grown as garden plants in Europe and elsewhere. What excited Charles Darwin so much is the unusual way in which the behavior and structure of the specialized moths that pollinate yuccas are adapted to their role in pollination of the flower. In other plant species, the insects that visit flowers are rewarded with a resource, often nectar, but the transfer of pollen between flowers, which is how the plant benefits from the transaction, is incidental to the behavior of the insect. Visiting insects become dusted with pollen, but they will often groom themselves to try to remove it.

In contrast, a female yucca moth has mouthparts with a unique tentacle-like structure that she uses to collect yucca pollen that she then stores in a batch under her head. The stored pollen can reach 10 percent of a female moth's body weight.

A female visits newly opened yucca flowers and if the flowers have not already been visited, the moth lays eggs in the ovary of the flower. The moth uses her mouthparts to scrape off a small sample of pollen from her store, then walks up to the stigma and deposits the pollen on its receptive surface with a series of bobbing movements. The sequence of egg-laying and pollination behavior may be repeated many times on a flower.

Yuccas and yucca moths are mutually dependent upon each other in an arrangement that could be characterized as: you scratch my stigma and I'll hatch your eggs. Trading a proportion of its seeds to feed moth larvae in exchange for pollination by female moths is a risky strategy for the yucca and, sure enough, some moths cheat. The cheats lay their eggs in fruit that have already been pollinated and have started to develop, thus exploiting both the pollinating moths and the yuccas. The advantage of this strategy from the cheaters' point of view is that if an ovary is too heavily parasitized by pollinators, yuccas will abort the fruit at an early stage. By arriving a few days after pollination, nonpollinating cheats can lay their eggs after the risk of fruit abortion has passed. The cost to the yucca is severe because cheats can triple the numbers of seeds lost to moth larvae. There seem to be about as many species of cheating yucca moths as there are pollinating species. Some cheating species have evolved from yucca-pollinating ancestors, while others belong to a group that has evolved alongside the pollinators.

The pioneers of pollination biology were frequently accused of telling tall tales, as Darwin found to his cost with the Madagascar star orchid. Several entomologists challenged Charles Riley's observations of yucca pollination, including one who claimed that it belonged "in the land of fables." Another, who was the first to observe a cheating species laying its eggs, mistook the cheating moth for a supposed pollinator and used the observation (which of course did not involve pollination) to challenge the veracity of Riley's description of yucca moths. In a paper that sorted out the confusion, Riley had his revenge by erecting a new genus for the cheating moths that he called *Prodoxus*, which is derived from the Greek for "judging something before having any experience of it." Through happenstance and the conventions of zoological naming, the entire group of moths to which yucca pollinators and cheats all belong is now known as the Prodoxidae. Let that be a warning of how the penalty for rushing to judgment can escalate.

Charles Riley was an early and enthusiastic Darwinist and was responsible for what was quite possibly the very first application of the new theory of evolution to solving an economic problem. In the early 1870s the French wine industry was on its knees due to the accidental introduction and spread of the grape phylloxera aphid from North America. The aphid feeds on the roots of vines, and Riley reasoned that North American vine species would be more resistant to the native phylloxera aphid than were European vines that had evolved in isolation from it.

When a population is exposed to a natural enemy such as phylloxera, more resistant individuals have an advantage because they survive better and leave more offspring than do susceptible individuals. This is natural selection at work: by this means, genes for resistance spread and over the generations resistance evolves greater and greater strength. Thus, Riley expected natural selection to have conferred resistance to phylloxera on North American vines that had long been exposed to the aphid. Conversely, European vines were susceptible because they had not been subject to natural selection for resistance through exposure to phylloxera before. Riley was right in his application of the theory of evolution by natural selection, and as a consequence the French wine industry was resurrected by grafting European grape varieties onto imported North American rootstock that was resistant to phylloxera. Charles Valentine Riley was appointed to the distinguished French Legion of Honor for this insight.

Since the unusual relationship between yuccas and their pollinators was discovered in the 1870s, four more examples of the phenomenon have come to light, each one having evolved quite independently from all the others. In Asia a group of shrubs in the Euphorbia family is pollinated by seed-parasitic moths; in Europe, globeflowers (*Trollius* spp.) belonging to the buttercup family are pollinated by flies that breed in their flowers; in the Sonoran desert the senita cactus (*Lophocereus schottii*) is pollinated by a parasitic moth. The number of plant species in all of these groups is relatively few, but there is also one highly successful and numerous plant group that is pollinated exclusively by seed parasites: the figs.

There are about 750 species of figs (*Ficus* spp.), so many, in fact, that they outnumber all other species in the mulberry family (Moraceae) to which they belong. While in the other known examples of pollination by seed parasites, like the yuccas, there are so few species that they can be dismissed as curiosities, in figs this method of pollination has become big business. Figs occur throughout the Old and New World tropics, in Australia, and in the Mediterranean. Their fruit is an important food resource for birds and mammals in many tropical forests.

The ancestors of the Moraceae had wind-pollinated flowers. Pollination by the tiny wasps that parasitize their flowers originated at least 60 million years ago, and since then the wasps and the figs have coevolved in a relationship that has driven the production of new species in both plant and pollinator. Figs have been unusually successful with a pollination mechanism that normally carries the ever-present danger that pollinators will consume too many seeds. The secret of this success appears to be that the plants are in control.

The edible fig is not a fruit in the strict sense, but actually a fleshy, unopened flower head with the flowers all on the inside. This receptacle is unique to figs. As one would expect in so large a genus, there is a good deal of variation among fig species in the details of pollination, but all share two features that are vital to the success of being a fig. Firstly, every fig species has its own specialized pollinating wasp species that are entirely dependent upon it for reproduction. This is a necessary condition for close coevolution between plant and pollinator and occurs in yuccas and other groups too. However, in figs the need for specialization by the pollinator is made especially strong because fig flowers are held within a receptacle that is closed to most insects.

Second, figs produce neuter flowers that provide brood chambers for their pollinators. Female fig wasps lay their eggs in neuter flowers that are incapable of producing seeds, but if they lay an egg in a female flower the egg does not develop and the seed is unharmed. Thus, the fig and not the wasp controls the resources devoted to pollination by fixing the number of neuter flowers in a receptacle. Though wasp larvae do not directly consume fertilized seeds as they do in yuccas, the neuter flowers in which they develop do represent resources and occupy space in the receptacle that a plant could otherwise devote to a female flower that would produce a seed. So figs have to pay indirectly for pollinator services with seeds, just like yuccas do, with the important difference that the plant and not the pollinator determines what the price shall be.

Fig species differ in how they combine the three different flower genders (male, female, and neuter) within a receptacle. Depending upon the species, a fig tree may produce a single kind of receptacle that contains all three flower genders; two receptacle types, one with female flowers only and the other containing neuter and male flowers; or three or four receptacle types with various combinations of sexual and neuter flowers in them.

Male fig wasps are wingless and live out the entirety of their brief lives within the receptacle where they are born, hatching, mating, and dying inside the fig. Female fig wasps leave the receptacle by a tiny pore at its apex, picking up pollen en route. Some species have been observed to store pollen grains away on their bodies, just as female yucca moths do. They then fly off in search of fresh receptacles in which to lay their eggs, pollinating the female flowers within when they do so. Unpollinated receptacles are shed by the plant, so any female fig wasp that neglected to repay her host by providing pollination service would be penalized by losing her offspring. One might imagine that a tiny female fig wasp only one or two millimeters long that lives for only two to three days would not be able to fly very far and that it would provide the fig with an inefficient means of outcrossing. However, paternity testing of fig seeds has shown that some are fathered by pollen that has traveled ten kilometers or more. Clouds of fig wasps from thousands of receptacles must be swept long distances on the wind.

The amazing relationship between figs and fig wasps was not discovered during Darwin's lifetime, but one can imagine how in awe he would have been of a coevolved pollination system even more sophisticated than that of the yuccas. The secret of success for figs must lie in the way that they have tamed their pollinators, farming fig wasps in neuter flowers in order to protect their precious seeds. The ability to pick and mix flowers of different gender within a receptacle confers flexibility on the fig reproductive system, which has enabled figs to adapt to a seasonal environment outside the tropics where they originated. The common edible fig *Ficus carica* is a native of the Mediterranean. At the end of the season wild populations of the species produce receptacles that contain only neuter flowers. Fig wasps are nurtured through the winter in these and then released to do their pollination duty by visiting fertile receptacles produced in spring.

Every coevolved relationship is susceptible to subversion, and the fig–fig wasp system is no exception. There is a whole community inside wild figs, including cheating wasps that breed in fig receptacles without pollinating them, seed parasites, and parasites of fig wasps. Humans, of course, exploit figs too, but we have domesticated varieties that produce seeds and sweet fruit without fertilization. There are probably no dead fig wasps in your dried figs, but you will notice plenty of nutritious, crunchy seeds.

{ 5 }

According to Their Own Kinds

INHERITANCE

The earth brought forth vegetation,
plants yielding seed according to their own kinds,
and trees bearing fruit in which is their seed,
according to its kind. And God saw that it was good.
GENESIS 1:12

Like begets like and thus differences between species are preserved, or so runs the biblical account. We owe a different view of inheritance to a Moravian monk, Father Gregor Mendel, who in the 1850s was fond of teasing visitors to the Augustinian monastery in the town of Brunn, capital of Moravia, by telling them, quite out of the blue and with a straight face, "Now I am going to show you my children!" While the visitors no doubt prepared themselves for the embarrassment of meeting the numerous illegitimate offspring of a supposedly celibate priest, Mendel would lead them with a smile into a small walled garden. Mendel's biographer, Hugo Iltis, describes what greeted them there: "Here there were to be seen, clinging to staves, the branches of trees, and stretched strings, hundreds of pea-plants of the most various kinds, with white and with violet blossoms, both tall and dwarf, some destined to bear smooth and others wrinkled peas."

If you had happened to glance from the monastery library window that overlooked this garden on a fine May morning, you might catch Mendel at work, tending his surrogate children:

The gardener would move from one flower to another, opening with fine forceps the blossoms that had not opened spontaneously, removing the keel, and carefully detaching the anthers. Then with a camel-hair pencil he would dust the pollen upon the stigma of another plant and would subsequently enwrap the flower thus treated in a little bag of paper or calico, to prevent any industrious bee or enterprising pea-weevil from transferring pollen from some other flower to the stigma thus treated, and in this way invalidating the result of the hybridization experiment.

(If Mendel had really wanted to shock his visitors, he would have told them he was going to show them his hybrids, since the German word for "hybrid" is *Bastard*.)

The lowly pea has an exalted place in the history of genetics, thanks to Mendel and to a peculiarity of its seeds which make them ideal material for the study of inheritance. Seeds are embryo plants swaddled, like overprotected infants, in layers provided by the mother that protect and control her offspring. If baby prefers the color blue, but mum likes pink, baby will wear pink. Mom decides, but it is not always so in the pea. To be sure, the pea pod is made by mother and her genes determine what color and shape it will be, and likewise the outermost skin that surrounds each pea in the pod. In some strains of pea this skin is translucent, and then some of the inherited features of the embryo pea plant within show through: sometimes the embryo is yellow, sometimes green. The shape of the pea, whether it is wrinkled or smooth, is another visible characteristic of the embryo.

You can tell that pea color and shape express the genetic characteristics of the embryo rather than just those of the mother plant because, contrary to the aphorism "as alike as two peas in a pod," peas in a pod are not always alike. The reason that peas in a pod may differ is that each one is the result of a separate union between an egg cell and the contents of a pollen grain, which means that they can be as similar, or as different from one another, as any siblings are. Dad's as well as mom's genes determine whether their little sweet-pea will be yellow or green, smooth or wrinkled. Characteristics that show in the seed simplify the practical study of inheritance because the results of crosses made between two different plants appear as soon as the seeds have ripened—there is no need to germinate and raise a whole new generation to discover what the results of the cross have been, as there would be in a study of the inheritance of flower color, for example.

The color of flowers interested Mendel greatly, and one of his *Fuchsia* hybrids with particularly good flowers was even marketed bearing his name to appreciative gardeners of the day. In the introductory section of the scientific paper about his researches with peas, which was eventually to make his name a household word, though only posthumously, Mendel wrote, "Artificial fertilizations undertaken in ornamental plants in order to produce new color varieties were the occasion of the experiments here described." Though Mendel possessed the practical and aesthetic impetus of the gardener, he clearly had his sights set upon obtaining the principles that governed inheritance, and as a means to this end he appreciated the experimental advantages of characteristics that expressed themselves in the seed.

What Mendel discovered in his garden was that when he fertilized the flowers of a plant grown from a wrinkled pea with pollen taken from a plant grown from a round one, all the resulting peas were round. Doing the cross the other way around, pollinating a plant grown from a round seed with pollen from a plant grown from a wrinkled one gave the same result: round peas. The wrinkled character had disappeared and all these peas *were* as alike as the proverbial peas in a pod. However, the picture changed completely when the seeds produced by the first cross were raised and fertilized with their own pollen. The seeds produced were now a mixture of round and wrinkled in a very particular ratio: three round to each wrinkled. What was extraordinary about these results was that a character, wrinkled seed shape, which disappeared from the progeny of the first cross had reappeared among the progeny of the subsequent one. This is rather like a magician's trick in which objects are made to disappear, only to reappear again with the flourish of a silk handkerchief. Where did the wrinkled peas go? The illusionist relies upon sleight of hand to hide objects, and Mendel concluded that nature must be pulling a similar trick, hiding the wrinkled trait. He proposed that there must be some kind of hereditary factor for seed shape which could be transmitted, but remain hidden, and then re-emerge again in a later generation. In finding evidence for such a hereditary factor, Mendel had discovered the gene, though that term was not coined for another forty years.

Mendel was not the first to discover that characters present in the parental generation, called the P1, could disappear in the next one (the first filial generation, or the F1), only to reappear again (in the F2) when these plants were self-fertilized (fig. 5.1). However, he was the first to realize what this meant, and he continued the experiment through further generations in order to test his ideas. The key to Mendel's breakthrough was that he recorded the ratios of the different characteristics that appeared in each generation, because these enabled him to work out what was going on. Self-fertilized flowers from the wrinkled peas produced by the F2 generation produced only wrinkled seeds, but self-fertilized plants grown from the round F2s fell into two

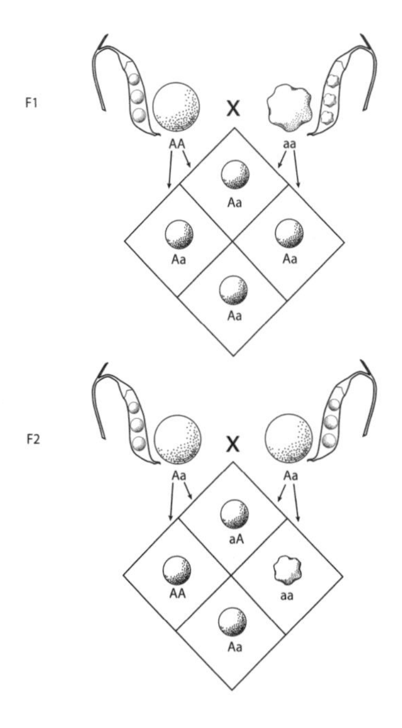

FIGURE 5.1. Segregation of seed traits in two generations of crosses between peas grown from round and wrinkled peas.

groups. One third of the plants produced only round seeds and two thirds produced a mixture of round and wrinkled. These mixtures were in the familiar ratio of three round to one wrinkled. Mendel's hybridization experiments with peas of different color (yellow or green) and with five other plant characteristics all gave the same pattern, with one type disappearing in the F1 generation and then reappearing in a consistent 3:1 ratio in the F2.

What did these consistent ratios, appearing and reappearing so faithfully, mean? Mendel concluded that in each case one characteristic, or trait, was dominant and its alternate form was recessive. The dominant trait, for example round seeds, was the only one observed in the F1, while the recessive wrinkled trait was the one hidden in that generation. Mendel proposed that there were in fact three types of plant produced in his crosses: pure-breeding dominants, which we nowadays denote AA; pure-breeding recessives, denoted aa; and hybrids, which are Aa. Any plant possessing a dominant A, whether or not a was also present, would show the dominant trait of round seeds. This simple classification brilliantly accounts for all of Mendel's ratios if we assume that the two factors which each plant carries are segregated from each other when pollen and ovules are formed, with one or the other present in each sex cell. Much later this idea became codified as Mendel's law of segregation.

The elegant way in which Mendel's system predicts the ratios of the seed types he observed can be seen in figure 5.1. The original crosses in the P1 were all between a round (AA) parent and a wrinkled (aa) one and thus all their offspring were Aa (round). When the F1 seeds were self-pollinated, segregation of the two factors followed by their random pairing in the fertilized seeds of the F2 produced the ratio AA:2Aa:aa. Since both AA and Aa contained the factor for the dominant trait of round seeds, the ratio of round to wrinkled seeds in the F2 was 3:1. When self-fertilizing the F2s, one third of the rounds, those with the AA constitution, bred completely true; two thirds of rounds (the Aa's) produced round and wrinkled seeds in the ratio 3:1; and the wrinkleds (all aa) produced only wrinkled progeny.

These results, repeated with seven different characteristics altogether, involved Mendel and his assistants in raising thousands of plants, pollinating tens of thousands of flowers, and counting hundreds of thousands of peas. There is a story that Mendel put his surplus peas to good use, keeping a pocketful to use as ammunition against sleeping pupils in the classes he taught at Brunn Modern School, but his former pupils remembered him with affection.

Most laborious of all the experiments were those Mendel performed to find out what the ratio of traits would be when he crossed plants with two or even three variable characteristics: for example round/wrinkled and yellow/green colored seeds. These experiments revealed that the ratios for such multiple traits were exactly those you would expect if the inheritance of seed shape and seed color were entirely independent of one another. This later became known as Mendel's law of the independent assortment of characters.

Today, when the gene is no longer just a convenient form of notation for explaining crosses, but has actually been pinned down to a thing with a chemical formula (deoxyribonucleic acid, or DNA), it is very difficult to grasp what Mendel understood his hereditary factors to be. Perhaps the A's and a's were just algebra to him, but whatever they were to Mendel, they were apparently all Greek to his contemporaries. Though rather, to be historically accurate, if an understanding of Greek had been all that was required to appreciate the significance of Mendel's discoveries, they would probably have received the attention they deserved in his own lifetime. Instead, they were ignored. Mendel's paper on his pea crosses, written in German, which was the principal language of science at the time, was published in 1866 in the *Proceedings of the Brunn Society for the Study of Natural Science.* Despite the parochial-sounding name of this journal, it was distributed to scientific libraries all over Europe, but Mendel's paper made not a ripple upon the surface of nineteenth-century science. When recognition came, like a storm from a clear sky, Mendel had been dead for sixteen years and his work had been neglected for nearly forty.

In the year 1900, in a triple coincidence that is now famed in the annals of science, a German, Karl Correns, a Dutchman named Hugo de Vries, and a Viennese, Erich von Tschermak, each independently rediscovered Mendel's work. First to announce his own experimental results in print was de Vries, whose paper arrived on Karl Correns's doormat one Saturday morning in April 1900. The contents of de Vries's paper hit Correns like a sledgehammer because he too had been carrying out crosses and he had obtained results like those of de Vries. The blow was doubly hard because de Vries and Correns were rivals of old and de Vries had pipped Correns to the post before. Correns had read Mendel's work and was dismayed to find that de Vries used Mendel's terminology of "dominant" and "recessive" traits, but did not mention Mendel at all. Correns wrote up his own experimental results and by Sunday evening had mailed them to the foremost botanical journal in Germany. In retribution for the suspected lack of candor by de Vries, Correns entitled his own paper "*G. Mendel's Law Concerning the Behaviour of Progeny of Varietal Hybrids*." This made it quite clear that if Correns himself was not the first to discover the laws of inheritance, neither was his rival from Amsterdam, de Vries. Last into the battle to inherit the mantle of the deceased Moravian monk was Erich von Tschermak. His paper, published a couple of months after the others, referred to Mendel, but showed little sign that he understood the true significance of the monk's work.

Long buried in the proceedings of a small Central European scientific debating society, when Mendel's paper finally emerged from obscurity it created plenty of controversy. The crucial question was, how general were Mendel's laws? Were they in fact laws at all? There were plenty of characteristics of animals, plants, and humans that did not seem to conform, with hybrids looking like a blend of their parents. The genes seemed only rarely to perform the illusionists' trick of the disappearing trait. Where was the magic in Mendelism? Was it just a sideshow?

The study of these questions brought a new science into being, the science of genetics. We now know that genes are the vehicles of virtually all hereditary traits, but that not all show dominance and recessiveness and therefore behave like the traits studied by Mendel. Many characteristics also rely on several-to-many genes, not just one, although the number of Mendelian traits in humans that are of importance to our health runs into thousands. Red-green color blindness is one, and favism, an acute intolerance of fava beans, is another. The importance of Mendel's discovery to modern science is difficult to overestimate. It is the foundation stone of genetics and therefore of the many aspects of medical science that depend upon a knowledge of genetics in humans, or in the viruses and bacteria that infect us. Without genetics there would be no real understanding of evolution and only limited progress in agriculture. Power to the pea!

After he completed his highly successful experiments crossing peas, Mendel wasted many years of his life futilely trying to make crosses between hawkweeds in the genus *Hieracium*, which, unknown to him, belong to that deceptive group of apomictic plants that require pollination to set seed, but produce fatherless offspring identical to mom. Progress in science depends so much upon having the right experimental system. Who knows how much further Mendel might have progressed if he hadn't been stymied by the bad advice of the senior scientist who recommended he investigate *Hieracium*? He might even have followed up on his investigations with corn, which he had found gave results similar to those of peas.

More than a century later, corn was the plant that earned the geneticist Barbara McClintock a Nobel Prize for her discovery that some genes just don't obey Mendel's laws. As an experimental subject for genetics, corn has the same advantage as the pea, which is that the result of a cross for a character like the color of the grain shows up in the seed itself. Better still, a corncob has perhaps a hundred seeds, where a pea pod has maybe six. Whereas pea genetics trades retail, corn trades wholesale, which means that rare events can be picked up much more easily in the results of corn crosses.

McClintock famously said that she had a "feeling for the organism" of corn. This gave her the ability to spot when something highly unusual and significant turned up among her thousands of corn crosses. What she discovered was that some genes don't follow the choreography of the chromosomes, segregating

{ 6 }

O Rose, Thou Art Sick!

ENEMIES

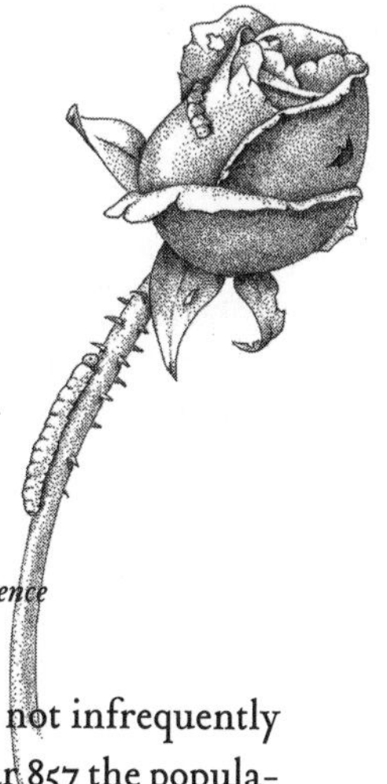

O Rose, thou art sick!
The invisible worm
That flies in the night,
In the howling storm,
Has found out thy bed
Of crimson joy:
And his dark secret love
Does thy life destroy

WILLIAM BLAKE, "THE SICK ROSE," FROM *Songs of Experience*

Every plant is plagued by natural enemies and not infrequently the same plagues destroy human life. In the year 857 the population of the small German town of Xanten, on the lower reaches of the Rhine, was struck by Divine Wrath: "A great plague of swollen blisters consumed the people by a loathsome rot, so that their limbs were loosened and fell off before death." In 994, Aquitaine in Southern France was struck and over forty thousand people died there. The approach of the first millennium and a belief that the world was coming to an end increased the terror. Holy water was sprinkled, and the bones of St. Martial, who it was believed had witnessed the resurrection of Christ, were disinterred to be used in sanctification. Forty years later there was an outbreak in Lorraine that was blamed on the pugnacious, who had broken the Truce of God that allowed fighting only Monday to Wednesday. Epidemics of the Divine Wrath were so frequent in continental Europe in the middle ages that the Order of St. Anthony opened hospitals specifically devoted to caring for sufferers, and the disease became known as the Fire of St. Anthony. The Abbey of St. Anthony at Viennes, near Lyon in central France, is said to have preserved the mummified limbs of victims until the beginning of the eighteenth century.

The cause of St. Anthony's Fire was poisoning by a fungus called ergot (*Claviceps purpurea*) that affects the seeds of wild grasses and cereal grains, particularly rye (*Secale cereale*). At least 132 epidemics of ergotism were recorded in Europe between 591 and 1789. Ergot is a sexually transmitted disease of grasses, dispersing its spores on the wind and infecting their promiscuously wind-pollinated flowers. Contact between sexual partners, even if the fluid connecting them is only air, is a channel of transmission often hijacked by parasites (think of HIV, Chlamydia, syphilis).

Transport by wind is a hit-and-miss business, but once the first infection has been achieved, the ergot fungus deploys something more effective by borrowing a trick from insect-pollinated plants. The infected flower is filled with honeydew that is stolen from the host's sap and laced with ergot spores. Nectar-seeking insects then transmit the disease to uninfected flowers in their search for further hidden filling stations. At a later stage the ergot fungus develops in place of the seed into a plump fungal body. Some of these are shed and sit dormant in the soil, like a malign mimic of a seed, germinating to release spores in a new season. Others may be gathered and threshed with the harvest and find their way into rye bread, turning the staff of life into a scepter of death.

Ergot contains a variety of toxic alkaloids, a large group of molecules that are notable for their powerful effects on the nervous system. The main ergot alkaloid is lysergic acid, from which LSD is manufactured. Another, ergotamine, is used in the treatment of migraine. Different strains of the ergot fungus produce different alkaloids, resulting in two distinct forms of ergotism. The victims of St. Anthony's Fire, whose dismembered limbs were preserved in the abbey at Viennes, were afflicted by gangrenous ergotism. Convulsive ergotism is not quite as gruesome, but has a long list of neurological symptoms and can also be fatal. Oliver Cromwell may have died of this form of ergot poisoning, suffering typical symptoms of insomnia, pain in the bowels and back, and fits. His death in September 1658 coincided with an epidemic of convulsive symptoms in England and occurred at a time of year when the rye harvest would have begun to enter the food supply. It was not till a century later that the cause of ergotism was recognized by doctors.

Before modern controls were introduced, ergot infection of rye was especially severe when abnormally cold winter weather conditions weakened the overwintering crop and the following spring season was especially damp, favoring the development and spread of the fungus. Just such conditions prevailed in the year Oliver Cromwell died. On the other side of the Atlantic a similar pattern of weather affected New England in 1691, and in early 1692 a series of events began to unfold that led to the notorious witchcraft trials in Salem, Massachusetts. In January 1692, nine-year-old Betty Parris, daughter of the Reverend Samuel Parris, and his niece, eleven-year-old Abigail Williams, began behaving strangely. They screamed, suffered contortions, uttered strange sounds, and complained that their skin felt as though it were being pricked with pins. The local doctor was consulted and when he could find no physical cause for the symptoms he suggested that the girls were the victims of witchcraft. Others fell ill with similar symptoms, cattle died suddenly after showing strange behavior, and a witch hunt began. By the time the governor of

Massachusetts called a halt to the trials, fourteen women and six men accused of witchcraft had been executed, including one man who was slowly crushed to death beneath stones in a futile attempt to force him to confess.

The evidence that the events in Salem and surrounding villages were caused by ergotism is circumstantial, but strong. The symptoms fit those of the convulsive form of the disease, the climatic conditions in which ergot prospers were present, and the households in Salem that suffered most were all located near soils well-suited to rye cultivation. Ergot infects wild grasses too, which may explain how cattle in Salem were also poisoned. Perhaps most convincing of all is the evidence of three women who attended a religious service given by the Reverend Parris, father of one of the first victims. The women declared that the bread used in the sacrament was red. Ergot contamination greater than 5 percent turns bread made with refined rye flour just that color. William Blake's words might be recast as an epitaph for this bitter experience: an invisible worm that flew in the night created a bread of crimson grief from which the superstitious villagers of Salem made a howling storm that did many lives destroy.

Though most of the many fungi that infect seeds are poisonous, there is one that that the Mexicans, culinary wizards to whom we owe corn, beans, squashes, chocolate, and chilies, consider a delicacy. Huitlacoche, or the Mexican corn smut (*Ustilago maydis*), is a fungus that commonly infects corncobs, producing a spectacular bulbous fruiting body that is used in mole sauces and as a vegetable. Like ergot, the corn smut infects its host via the flowers, taking the route normally used by pollen to reach the ovules by growing along the length of the corn silks that hang from the cob. If a pollen tube gets to the ovule first, the silk route is closed to infection by corn smut and at least some seeds in the cob escape infection.

The grass family, to which corn and rye belong, is unusual among plants in having practically no chemical defenses against the many animals that would like to lunch on them. Pasture grasses survive the attentions of cattle, sheep, and horses only because they can regrow lost leaf tissue from buds that are buried beneath the reach of these herbivores' jaws, but even these grasses suffer if grazing is too frequent. This is where a group of poisonous fungi, which began life as parasites inside grasses, can prove themselves useful to their otherwise chemically defenseless hosts by producing alkaloids that deter or even kill would-be grass eaters. The so-called "endophytic" fungi live inside stems, leaves, and even in some cases the seeds of grasses. Perennial ryegrass pastures in New Zealand and elsewhere are heavily infected with endophytic fungi that poison livestock and produce a cattle disease known as "staggers." The alkaloids responsible for staggers are very similar to those produced by ergot.

In the most widespread group of endophytes, belonging to the genus *Epichloë*, different species play different roles, from a villainous enemy of grasses which they render sexless and seedless, to a constant companion that protects the host from insects and drought and secrets itself away in the embryo of the seed, providing a providential inheritance for the grass's offspring. Plants that are infected by benign, sexless endophytes show no external symptoms of endophyte infection because the fungus never produces spores. Other species of *Epichloë* seem sometimes to be friends and sometimes foes of grasses, depending upon ecological circumstances. Many intimate biological relationships in nature are similarly ambivalent and fickle.

Curiously, the *Epichloë* species that do go in for sexual reproduction rely upon a parasitic fly called *Botanophila* to achieve sexual union between compatible spores. The fly feeds upon a fungus, ingesting spores in the process. When it flies to another infected plant, it lays its eggs and then cross-fertilizes the fungus by dragging its abdomen over the surface in a stereotypical manner, smearing it with feces and spores that pass intact through the fly's gut. *Botanophila* larvae are able to develop only on a fungus that has been cross-fertilized, thus making fungus and fly dependent upon each other for reproduction. The parallel with the moths that simultaneously pollinate and parasitize the seeds of yuccas is uncanny.

Evolution in *Epichloë* seems to be actively exploring the full gamut of grass-fungal relations. Out of the seeming chaos that rages between fungal amity and fungal enmity a constant evolutionary law emerges: the endophytes that are plant-friendly give up sexual reproduction and are transmitted via grass seeds; the endophytes that are plant-unfriendly and that neuter their hosts reproduce sexually. It's almost as though endophyte and grass are constrained to share sex in the same way that the three Graeae of Greek mythology shared a single eye and one tooth between them, passing them from one to the other. Only one may possess the facility of sex at any one time. Indeed, to the fanciful imagination, fungal-seed relationships possess all the hallmarks of witchcraft: invisible allies, mortal enemies, pricking with pins, poisons and mind-bending drugs, sex and castration. Consider, by comparison, the ingredients of the famous witches' brew in *Macbeth*:

Eye of newt, and toe of frog,
Wool of bat, and tongue of dog,
Adder's fork, and blind-worm's sting,
Lizard's leg, and howlet's wing,
For a charm of powerful trouble,
Like a hell-broth boil and bubble.

You wouldn't want to taste the concoction, but it probably wouldn't do you any harm. How much more potent and deadly a seed-based recipe would be—but who is going to be scared by a vegetarian witch's granola?

Smut of rye in daily bread
with peanut butter thickly spread.

Smother it with castor bean,
add strychnine seeds to cause a scream.
For a charm of powerful weeds,
look no further than these seeds.

Castor beans contain ricin, the most deadly poison known. For extra potency, the peanut butter should be made from nuts infected with *Aspergillus*, a fungus that produces cancer-causing aflatoxin. In truth, those accused of witchcraft in the middle ages were frequently healers learned in wortcunning—the knowledge of herbal remedies. They would have been much more familiar with the powers of seeds, for good or evil, than with how to obtain a blind worm's sting.

About six hundred different species of fungi are known to infect seeds and to use this route to propagate themselves. These fungi include not only endophytes that may be benign or even beneficial to the plant, but also many pathogenic species, such as the fungus that causes the devastating wheat disease called bunt. This fungus destroys the flowers of infected plants, and then spreads by spores to uninfected plants, using the seeds of these to transmit itself to the next generation. The bunt fungus thus alternates its mode of transmission, obtaining the best, or from the perspective of the plant, the worst, of both worlds.

Not only fungi, but also bacteria, viruses, and insects exploit seeds as a combination of vehicle and lunch wagon during critical phases of their life cycles. Among insects the chalcid wasps, the group to which fig wasps belong, are specialist seed parasites that pass the majority of their lives as larvae or pupae inside seeds. Seeds of Douglas fir *Pseudotsuga menziesii* are infested by a chalcid that is especially well adapted to this way of life. In any relationship involving reproduction, timing is always critical. To father a seed, pollen must be shed when female cones are receptive, and to parasitize that seed a chalcid wasp must lay her egg at the critical moment. For most seed parasites this means delaying egg laying till the ovule has been fertilized, because only then can a food supply for the larva be guaranteed. Unfertilized ovules are normally aborted by plants. This is why cheating yucca moths delay egg laying.

In conifers there is a long delay between pollination (the arrival of pollen on the receptive surface of the stigma) and fertilization (the fusion of sperm and egg nuclei). In Douglas fir this delay can be up to ten weeks, but the chalcids that parasitize its seeds do not wait for fertilization to take place and seem to lay their eggs indiscriminately. It turns out that these chalcids can ignore the risk of seed abortion because, if an egg is laid in an unfertilized seed, the insect performs a hormonal deception that makes the plant provision the seed and its parasitic occupant, even though no plant embryo is present.

In any conflict, whether between trees and seed predators, or between hostile nations, there are always alternative strategies that can be followed. Circumstances, choice, and chance dictate whether opponents will take the high road of peaceful, if wary, coexistence or the low road to escalation and war. In July 1959 Richard Nixon, then vice president, visited Moscow to open an exhibition showcasing American technological achievements. The Soviet Union had two years earlier launched *Sputnik*, the first orbital satellite, but Nixon pointed to the centerpiece of the show, a model American home with a kitchen fully equipped with electrical appliances. Showing the premier of the Soviet Union around the kitchen, Nixon asked Nikita Khrushchev, "Would it not be better to compete in the relative merits of washing machines than in the strength of rockets?" Nixon's suggestion was probably tongue-in-cheek, but anyhow Khrushchev clearly thought so. The ensuing three decades of U.S.-Soviet relations were dedicated not to peaceful competition in the technology of home appliances, but to a military arms race driven by a shared strategy of Mutually Assured Destruction.

Red crossbills (*Loxia curvirostra*) and lodgepole pines (*Pinus contorta*) are locked in an arms race. This has shaped the evolution of bird and tree, just as the cold war left its mark on the histories of the United States and Russia. The lodgepole pine protects its seeds with heavy armor, sealing them with resin inside woody cones. Crossbills are named for the way the upper and lower halves of the beak cross. Cross your fingers and you will create the same effect. This orthodontic nightmare enables crossbills to pry apart the scales of conifer cones to extract the seeds within. How can conifers cope with seed predators that come in flocks so well armed to steal their precious offspring from the arboreal cradle? The boughs need not even break for these particular babies to go missing from their treetops.

In the Rocky Mountains, top dog among seed predators is the pine squirrel *Tamiascurus hudsonensis* which strips whole trees of all their cones. Red crossbills cannot cope with this scale of competition for seeds, so the birds are rare in mountain valleys where pine squirrels are present. But in secluded valleys where there are no pine squirrels, red crossbills and lodgepole pines are eyeball-to-eyeball in a war over seeds. Here, the cones of lodgepole pines are longer, narrower, and heavier and have thicker scales than in valleys where crossbills are absent. Some of this difference may be due to the absence of selection by squirrels, but the presence of selection by crossbills is implicated too. Red crossbills prefer smaller, shorter cones with thin scales at the top of the cone. The cones of lodgepole pines have evolved in just the opposite direction wherever red crossbills are the main seed predator. In response, red crossbills have evolved a stouter beak than is found in other populations. This adaptation enables them to attack cones more easily, but at a cost. A larger beak comes with a larger body size that requires more food. For seeds too, size and food are a matter of survival.

{ 7 }

The Biggest Coconut I Ever See

SIZE

Tell me what you want to know
How tall a palm tree does grow?
What is the biggest coconut I ever see?
What is the average length of the leaf?
Don't expect me to be brief
cause palm tree history
is a long-long story
JOHN AGARD, FROM "PALM TREE KING"

Like some of the most spectacular fossil dinosaurs, the rare coco de mer palm (*Lodoicea maldivica*) was first found by the discovery of stray pieces of its anatomy. Huge "double-coconuts," the size of an overinflated basketball, were collected from the beaches of the Maldive Islands, and hence the species was given the epithet *maldivica*, even though the palm itself does not occur on those islands. Its true home turned out to be 2,300 kilometers southwest of the

Maldives, on two small islands in the Seychelles. The fruit of the coco de mer contains by far the biggest seed known, the whole thing weighing up to twenty-three kilograms when ripe. That's the same as the typical checked-baggage allowance you get when flying economy.

Though called a double coconut, the nut usually contains only one seed, but this has a pair of large, rotund lobes which bear a striking resemblance to female buttocks. In the sixteenth and seventeenth centuries this resemblance led people to expect medicinal and aphrodisiac properties of the coco de mer, and a single nut could fetch four hundred pounds in London, even after the source in the Seychelles had been discovered. The association with concupiscence has proved to be as enduring as sex itself, and in 2003 an up-market sex shop using the name Coco de Mer opened in Soho in London.

Why is the double coconut so big? The answer does not seem to have anything to do with buoyancy or dispersal. Fresh seeds of *L. maldivica* do not float, and those found washed up on distant beaches have never germinated, so the actual dispersal powers of the double coconut are nothing like those of its namesake *Cocos nucifera*, whose seafaring success is legendary. It would be more appropriate to call it coco de terre than coco de mer. Is there any other plant that is so misunderstood that it has three different misnomers in French, English (double coconut), and Latin (*L. maldivica*)?

Since all the double coconuts that get washed away from the Seychelles die, the giant size of this nut cannot be designed by evolution to travel, but to endow a seedling that stays at home with the nutrients it needs to survive in its native habitat. What is it about that habitat that demands such a huge seed, and how did it evolve? For so famous and extreme a monstrosity of nature, there has surprisingly been only one attempt to explain how it evolved, and that was published as recently as 2002. The following account is largely based upon a research paper by Peter

Edwards and two colleagues at the Swiss Federal Institute of Technology in Zurich.

One of the last two strongholds of the coco de mer in the Seychelles is on the island of Praslin, where the Vallée de Mai National Park in which it lives has the status of a World Heritage Site. The island is geologically ancient, consisting of a granite mountain that is a wayward fragment that probably became detached from India some 75 million years ago as the subcontinent broke away from the great southern continent of Gondwanaland and rafted northward. Most of the native plants found in the Seychelles today are therefore long-haul passengers from the ancient Gondwanaland flora, rather than recently arrived colonists of the kind that are found on newer, volcanic archipelagos such as Hawaii. The closest living relatives of the coco de mer are thought to be palms in the genus *Borassus* that are not found in the Seychelles at all, but in Asia and Africa, where they live in much drier habitats than the tropical forest where *Lodoicea* occurs. Thus the coco de mer is not only distant from its geographical origins, but also far from the kind of dry habitat to which its ancestors were probably adapted. How did evolution turn an unremarkable, *Borassus*-like palm of dry, savannah habitats, probably bearing fruit with several golf-ball-sized seeds in them, into a tree with single-seeded fruit as big as a bomb?

Peter Edwards and his colleagues suggested an ingenious answer to this question which neatly ties together all the evidence. As the Seychelles were carried northward in the wake of India, their climate would have become more moist, resulting in a gradual change from the dry kind of habitat to which *Borassus* might be suited to the wet tropical forest conditions of today. When climate changes on a continent, plants migrate across it into new terrain as it becomes habitable. Forest trees all across the Northern Hemisphere did precisely this as the glaciers retreated at the end of the last ice age. Seed-eating animals, such as pigeons and jays, are the vessels in which the migrating plants travel, and plants can progress a hundred kilometers in a generation by this means. However, when the climate changes on an oceanic island like Praslin there is very little scope for new habitats to be filled by a cast of preadapted species because there are none waiting in the wings and the isolation of the island makes it unlikely that any will be able to reach them from elsewhere. The absence of species already adapted to changing conditions gives species that are already present a chance to evolve to match the new demands and opportunities of the changing environment. The slow rate of continental drift which ensures that climatic change is relatively gradual might also have facilitated evolutionary change.

Forests require moisture and moisture conjures trees from the soil wherever their seeds are present. As the climate of the Seychelles became more moist and vegetation became taller, the seedlings of the proto-coco, as we might name these ancestors, would find themselves having to compete with taller and taller plants for light. The individuals with the biggest seeds nearly always win a contest like this and so, as the height and density of the vegetation increased with increasing moisture, natural selection would have favored plants with larger and larger seeds. Today, the seed of the double coconut fuels the growth of the seedling so effectively that even its first leaf has a stalk one and a half meters long, and a juvenile plant can thrust its leaves to a height of ten meters in just a few years.

Palms have an inherent disadvantage compared with other trees because they lack the ability to increase the girth of the trunk as they get taller, as oaks or pines do, for example. Because palms cannot do this, a juvenile must establish a trunk that is stout enough to support a fully grown tree decades before such support will actually be needed from an engineering point of view. It is as if every newlywed couple were forced to purchase, without a mortgage, at the time of their nuptials a large, family-sized house far too big for their immediate needs. The only way such a thing can be done is if your parents endow you with a large amount of cash or a fat seed. In seventeenth-century London, a single coco de mer seed would undoubtedly have fetched enough money to buy a handsome house!

So the need to establish in shade is how the proto-coco probably began its career in show business, but how did it end up topping the bill? This is where we must suspect another factor came into play. The bigger the nut of the proto-coco became, the nearer it fell to its parent tree. The worst place to try to establish yourself is right beneath mom, not just because she will always be bigger than you and cast you in heavy shade, but also because all your siblings will be there too and you will have to compete with them also. The solution that evolution came up with to this problem is stunningly simple: if the seed is too big to move, then move the seedling!

The coco de mer and its continental relatives germinate in a rather peculiar way. The seed sends out a kind of umbilical cord which buries itself in the soil and tunnels away from the seed at a depth of thirty to sixty centimeters beneath the soil surface. This is a trick that evolved long ago among the ancestors of *Borassus* and *Lodoicea*, but the coco de mer picked it up and ran with it—its cord, sometimes called a "rope," can reach up to ten meters in length before a seedling appears on the end of it. The rope is the seedling's pipeline to the food store in the seed, and the connection may persist for four years. This extraordinary mechanism allows the coco de mer to turn the large size of its seed, which is normally an impediment to dispersal, into an asset that actually helps in the dispersal of its offspring. By reversing the usual relationship between dispersal distance and seed size, *Lodoicea* has broken one of the constraints that normally limits the advantage of having big seeds—that poorly dispersed offspring end up competing with one another. Thanks to the rope, bigger means better all round for double coconuts, and thus evolution has driven them to an unprecedented size.

There are, of course, some significant costs involved in producing such colossal seeds. Development is protracted and it can take a decade for a nut to reach maturity. The number of seeds a female palm can produce is strictly limited, and most trees probably produce well under a hundred in a lifetime. A large palm may have up to half a ton of developing fruit in her crown, and the enormous weight is a handicap in high winds, which can decapitate female trees as a consequence. Most palms cannot branch and so decapitation is fatal to them. This is such a serious hazard for female coco de mer that the sex ratio of mature palms, which ought to be 50:50, is male-biased, with nearly two males to every female in the Vallée de Mai.

The coco de mer is a bizarre and wonderful monstrosity, but it has to play the game of evolution by the same rules as everyone else—what helps a plant leave more offspring will be favored by natural selection. The result has been so unusual in this particular palm only because its circumstances were so special and because its ancestors had evolved and passed down to it a rope trick that it could turn to its advantage. Big seeds were the result.

The range of seed sizes among plants in general is huge. The double coconut is about 20 billion times heavier than the smallest orchid seeds, some of which weigh only a ten-millionth of a gram. Seeds of this minute size are viable only because the tiny orchid seedling lives parasitically upon a fungus during its early years. Among the rest of the seed plants, a significant part of the variation in seed size is due to differences in growth form between species. Trees usually have bigger seeds than herbs, and this difference goes right back to the dawn of angiosperm evolution. The very first flowering plants were herbs or small woody plants and had smaller seeds than their gymnosperm ancestors, which, like modern gymnosperms, were probably woody trees or shrubs. This trend toward smaller seeds was reversed when the palms evolved from herbaceous ancestors. On average, the seeds of palms as a group are over four hundred times bigger than the seeds of the much smaller, herbaceous plants to which they are most closely related.

A big seed is clearly an advantage to a seedling that must, like the double coconut, survive strong competition from much taller plants, but why should any plant evolve smaller seeds? How can smaller seeds be an advantage? The answer is that each seed is like a ticket in a lottery for survival. The more tickets you have, the greater the chance that at least one will win. So if circumstances permit smaller seeds to survive, perhaps because competition from trees is checked by disturbance or some other environmental constraint, then natural selection will favor plants that produce many small seeds over those that produce just a few big ones. This is why, in the poetic words of Charles Darwin's grandfather, "Each pregnant oak ten thousand acorns forms."

{ 8 }

Ten Thousand Acorns

NUMBER

Each pregnant oak ten thousand acorns forms
Profusely scatter'd by autumnal storms.
ERASMUS DARWIN, FROM *The Temple of Nature*

In a good year, the sheer profusion of acorns is surely testament to the bounteousness of nature, particularly if you are a squirrel or a mouse. The value of this bounty was well recognized in ancient times, when in old English it was called *mæst*, meaning "forest food." From this is derived the modern term, perhaps now known only among foresters, of "mast." Though the food is free, mast tends to be an irregular feast, and most nut-bearing trees, like oaks, beeches, or pines, have distinct mast years with several years of dearth between the much rarer years of plenty. In the fall of a mast year, rodents wax fat on a high-carb diet and bury some of the surplus acorns in caches that they retrieve and eat later.

Gray squirrels can discriminate between weevil-infested

acorns, which they eat right away, and uninfested ones that will keep. Not only weevils, but germination could rob them of their cache before they return to it. Some squirrels ensure that their cache cannot germinate by nibbling out a notch from the tip of the nut before burial. This destroys the embryo in the seed, leaving the main food stores of the acorn intact. Certain individual oaks have evolved a countermeasure to this threat and produce acorns with a displaced embryo that is not damaged by nibbling. Squirrels notching acorns to destroy the embryo and oaks moving embryos to avoid this are just two bouts in the constant evolutionary battle between predator and prey.

Mice and squirrels are not the only guests at the mast-year banquet; deer and birds also join in the feast and in former times humans ate acorns too. Acorn remains have been found in excavations of the Neolithic town at Catal Huyuk, in Turkey. These indicate that eight thousand years ago, near the dawn of agriculture in the Fertile Crescent, acorns were an important part of the diet. They may even have been the first staple food to support permanent settlements, before wheat and barley were domesticated in the Fertile Crescent. A lot less labor would have been required to gather and store acorns than to sow and harvest cereals.

For thousands of years, until the late nineteenth century, acorns were also a staple for Native Americans. The Miwok tribe, who inhabited oak woodlands near the present-day town of Jackson, California, ground acorns into flour with pestles and mortars. The communal mortars that furnished their permanent settlement are still in evidence in a limestone outcrop pitted with over a thousand holes at Indian Grinding Rock State Historic Park. The Miwok stored acorns that they collected in mast years in granaries that sustained them through lean years. Before they can be eaten, most acorns have to be leached of tannins that make them bitter and indigestible. Some Californian tribes would bury their acorns at the edges of streams where the flow of water leached them naturally and the lack of oxygen preserved them from germination or decay. Acorn flour is still eaten in Korea and packets of it are sold in Korean grocery stores in the United States and Europe.

The influence of the giant pulse of food injected into forest ecosystems in a mast year ripples out through the food web and, for short-lived animals like mice and insects, it reaches across the generations. The cornucopia of acorns in a mast year is followed by a multitude of mice in the next. This is because in the winter of a mast year woodland mice survive the perils of the cold season in greater numbers than usual, and the plentiful survivors have more babies in the spring. But these baby boomers are born into a world with many more mouths and far less food than in their parents' generation, so they must turn to other sources for sustenance. A study in New York State showed that in the season after the mast year of 1994, densities of white-footed mice were up fifteen-fold. With the acorns all gone, mice turned to other food, like pupae of the gypsy moth, which overwinter in the soil; predation on pupae increased thirty-four-fold. The gypsy moth is an introduced pest that regularly defoliates vast tracts of forest in eastern North America. In 1995 the increased predation on gypsy moth pupae caused by the boom in mice spared the oaks from gypsy moth caterpillars that year.

Though a murine Dr. Pangloss born in the *annus mirabilis* of 1994 might conclude that the bounty of the mast, so marvelous for mice, deer, birds, and future oak leaves, was ample evidence that "all is for the best in the best of all possible worlds," he'd only have to be born a year later to think differently. Just as in the stock market, so for acorns, boom is followed by bust and the ecological repercussions are just as severe and widespread as the economic ones. In fact, the acorn boom itself has a downside, not just for gypsy moth born under a bad sign, but for humans living in the leafy neighborhoods of rural Connecticut too.

Rural New England today is as amply oaked as a California

Chardonnay, but these forests aren't ancient; they are about as recent as the California wine industry. The stone walls that once marked field boundaries can still be traced among the woods of New England. They are remnants of farming that became uneconomic when the more fertile and easily tilled land of the Midwest was opened for agriculture in the mid-nineteenth century. New England's fields were abandoned and the woods recolonized them. Given the explosion of the cities and suburbs that has taken place it may be hard to believe, but New England has more forest cover now than it did in the days when Thoreau sought solitude on Walden Pond. But not all is well in the woods.

In the 1970s a new, debilitating disease was uncovered in the rural community around Lyme, Connecticut. Patients reported skin rashes, fever, and chronic arthritic pain in the joints. Children and adults alike were affected. Whole families sickened and cases were clustered in a manner previously quite unknown for arthritis. In the early days, when its cause was still unidentified, one doctor investigating the disease reported "On some streets, it was just one house after another. . . . I had been given a list of names to telephone. Once I dialed the wrong number and got a home where that child had arthritis!" The etiology that was eventually unraveled for Lyme disease, as it became known, linked the incidence of the disease with acorn mast with a time lag of two years.

It turned out that the causative agent for Lyme disease was a type of bacterium called a "spirochete" (because of its corkscrew appearance under the microscope). The particular species of spirochete was new to science and was named *Borellia burgdorferi*. The spirochete infects both white-footed mice and white-tailed deer and is transmitted between the two animals by the black-legged deer tick, which spends a part of its life cycle on each host. Humans become infected by *B. burgdorferi* when bitten by an infected black-legged tick that is searching for a new host. Usually, but not always, a rash like a bull's-eye develops at the site of the tick bite. Even if not treated, the rash usually clears up, but the infection progresses and in later stages it leads to arthritis and sometimes severe neurological symptoms.

The life cycle of the deer tick, from egg to egg-laying adult takes two years. *B. burgdorferi* is not usually transmitted via the egg, so a newly hatched tick looking for its first host begins life free of *B. burgdorferi*. Without a reservoir of infection, the spirochete would disappear between one generation of ticks and the next. The reservoir of *B. burgdorferi* is in the population of white-footed mice, from which juvenile ticks acquire infection when they feed upon their blood.

Two factors increase tick densities two years after a mast year. Initially, the large acorn crop concentrates deer and mice together, increasing the likelihood of transmission of ticks between them. Subsequently, mouse density increases, providing more hosts for juvenile ticks. The most likely time for a human to catch Lyme disease is when frequenting the woods in the second season after a mast year (e.g., 1996) when ticks are leaving mice in search of a second host. Lyme disease is probably more common in the New England than formerly because the deer population has increased dramatically, helped by the food and shelter that the regenerated forests provide. With more deer about, more ticks complete their life cycle. Lyme disease also occurs in Europe, though it is less common there. With deer populations also exploding in parts of Britain, a rise in Lyme disease is a possibility, especially as more forests are planted.

The periodic bumper crops of seed produced in mast years are all the more important to animals because all trees of a species, and sometimes even different species, mast in synchrony over distances of up to 2,500 kilometers. In mast years there is food everywhere, but betweentimes there is nothing. When fat years are suddenly followed by lean, the result is an eruption of seed-eating birds from their usual habitats all over the Northern Hemisphere, with American species such as the black-capped chickadee and pine siskin spilling in large flocks beyond their normal ranges.

Why do oaks and so many other forest trees, in the tropics as well as the temperate zone, vary their annual seed production so drastically from year to year? After all, masting has some obvious disadvantages. First, any seed production is a physiological drain on a tree's resources, but a concentrated burst is enough to slow down growth for at least one or two years afterward. Second, of course, mast crops shout "come and get it!" to every seed-eating animal that wants a free lunch. Third, masting trees pass up potentially valuable opportunities to reproduce in the in-between years. At these times, offspring from nonmasting trees might establish without competition from seedlings belonging to the-crowds-that-mast. Quite clearly, masting trees have blundered, and the people who talk to plants haven't passed on this information. Or should we listen to the trees?

The simplest explanation for why trees mast would be that they just can't help it because they are forced to track variation in the weather. Although trees appear to use climate to trigger synchronized seed production, climatic events cue mast years rather than driving them. The difference between a cue and a driver of masting is analogous to what happens in a motor race. The checkered flag is the signal that starts the race, but it does not actually propel the cars forward. The speed of the cars is determined by the driver in each vehicle. In tropical Southeast Asia, where the climate is not seasonal, dipterocarp trees appear to use very slight changes in temperature related to El Niño to cue their mast years. They produce seeds only every seven or eight years, but when the flag goes down they really floor the accelerator and produce massive numbers. In Northern Hemisphere forests too, yearly seed crop variation is much more extreme than variation in the weather that cues it. So the idea that weather drives masting doesn't fly. Has nature blundered, or could masting have advantages?

Needless to say, nature has not blundered. On the contrary, masting is one of those extraordinary phenomena of nature that is quite inexplicable without natural selection. The answer is simple. Small seed crops are usually completely wiped out by seed-eaters, but a fair proportion of much larger crops survive, despite the hoards that are drawn to feed on them. The phenomenon is known as "predator satiation." Masting is a strategy to outwit seed-eaters by surfeiting them with food in mast years and then starving them between times. The effect that sudden crop failures have upon seed-eating birds is an example of how effective the strategy is.

Masting by trees has forced the evolution of countermeasures by animals. Rodents store seeds, but the brevity of their lives means that many caches may not be recovered. Unless disabled before burial, these seeds can germinate, and the rodent that carried them away has, in the end, performed a service of dispersal for the tree in a failed attempt to provide for itself. This is an example of how evolution continually subverts the strategies of one species to the ends of another. In this case, it has resulted in a kind of mutualism in which trees sacrifice a fraction of their seeds as a taxi fare in exchange for the dispersal of the remainder of their offspring.

Migration is another recourse for starved seed-eaters, but a desperate one since seed crops may be synchronized over thousands of kilometers and so seed famines can be very widespread. Among some seed-infesting insects, another strategy is to wait out the intermast years as a dormant pupa inside the seed that has been their cafeteria and nursery. A neat trick if you can manage it. Insects that cannot do this have a problem, as illustrated by a weevil that infests pecan nuts.

If you love pecan pie, as I do, then we should both thank another aficionado of this tasty nut, a moth called the pecan nut casebearer. Strangely, though the casebearer's caterpillars destroy pecan nuts in bud, masting turns these insects into allies and not enemies of the pecan tree. This is how it works: Pecan trees produce most of their crop in mast years when a small fraction of the crop is lost to the casebearer's caterpillars, but the chief consumer of developing nuts in a mast year is the pecan weevil. The next year there are more casebearers, lots and lots of pecan weevils, but a pecan famine. Pecan trees do produce a small number of flowers and nuts between mast years, but the casebearer prevents weevils laying eggs in nonmast years because it gets to the buds first and destroys them.

If the casebearer did not rob the weevil of a food supply between mast years, the weevil, boosted by the large numbers born in mast years, could potentially sustain itself at a high level by ticking over in the nuts produced between times. It would then be more difficult for the pecan tree to satiate its predators in a mast year and to allow a proportion of the nuts produced to escape. It has been calculated that by destroying potential breeding places for the pecan weevil in nonmast years, the casebearer moth saves an average pecan tree from an additional lifetime loss to the weevil of over 70,000 nuts. The cost of this protection to the pecan tree is a mere 200 nuts lost to the casebearer.

The example of the pecan and its competing seed predators illustrates how an understanding of ecological interactions can be turned to practical advantage. From the perspective of the pecan tree, and the pecan nut farmer, the right attitude to the casebearer caterpillar ought to be "my enemy's enemy is my friend." Every organism, including those we depend upon for our food, exists in a web of ecological interactions, and most of our enemies have enemies of their own. Gypsy moth pupae are attacked by the white-footed mouse, for example. Intelligent use of natural enemies to control crop pests and invasive species is known as "biological control." Biological control is becoming of increasing importance to agriculture and horticulture as the disadvantages and dangers of some chemical pesticides become clear.

The cost of seed dispersal by animals that oaks and other nut-bearing trees pay by sacrificing a fraction of their offspring is paid in another way by plants bearing fleshy fruits. In apples, cherries, pears, peaches, plums, and other fruit trees, each seed offers its own, individual taxi fare in the form of a succulent, usually sweet, wrapping that conceals the seeds. The dispersal strategy of fruit trees bears an interesting comparison to that of nut trees. Both types of tree use animals to disperse their seeds and for both the bargain between plant and animal is a treacherous one. "Look how rich and tasty I am; come and pick me up" is not how I would hail a Yellow cab on the streets of New York. How do you prevent a carrier taking more than the fare you bargained for?

The answer depends upon the way in which you pay the fare. Nut trees have a kind of advance-payment system, using a proportion of their seeds to purchase the dispersal of the remainder. Nuts must be palatable to animals for the system to work, but trees limit lifetime seed losses by masting and predator satiation. Fruit trees, on the other hand, purchase dispersal with a pay-as-you-go system. For fruit trees it would be counterproductive to satiate their dispersal agents, so they tend to produce regular crops and they do not mast. Instead, they often protect their seeds by making them toxic. Apple and peach seeds, for example, contain cyanide.

An apparent exception to the rule that, unlike fruit seeds, nuts are palatable is the cashew nut, which grows in a shell protected by a toxic lining. The lining of the shell contains a resinous oil that blisters the skin, like the worst case of poison ivy that you can imagine. Of course this oil is removed in processing before cashews go on sale. But in the wild, how does a nut that severely blisters the skin if you bite into its shell get dispersed? The answer is in a feature of the cashew nut that you would never guess from just seeing the kernel. Cashew nuts grow attached to a juicy fruit! Each mature nut hangs from the base of a pear-shaped, fleshy

{ 9 }

Luscious Clusters of the Vine

FRUIT

What wondrous life is this I lead!
Ripe apples drop about my head;
The luscious clusters of the vine
Upon my mouth do crush their wine;
The nectarene, and curious peach,
Into my hands themselves do reach;
Stumbling on melons, as I pass,
Ensnared with flowers, I fall on grass.
ANDREW MARVELL, FROM *The Garden*

In his solitary reverie Andrew Marvell imagines that nature is thrusting its fruits upon him, as if he were Adam in Eden before the fall. Three hundred and fifty years after Marvell wrote *The Garden*, even an unbeliever must agree with him that fruit seems to demand to be eaten. Is there an evolutionary explanation for this? Indeed there is, if we think about the question from a plant's point of view. A fruit is the package in which a plant's seeds are sent out into the world, often ornamented to attract animals and upholstered with tasty flesh to reward them. A fleshy fruit is the vehicle, the seed its cosseted passenger, and birds and mammals the motive power for dispersal. Berries are juicy baubles that bring birds flocking to carry away a plant's offspring. Some berry-bearing plants spread so successfully that they have become serious weeds.

Seed and fruit go together more naturally than either horse and carriage or love and marriage; they are biologically bound by a knot tied deep in the ancestry of seed plants. By contrast, from a linguistic perspective, the metaphors "seed" and "fruit" have an almost antithetical relationship with one another. The word "seed" connotes a beginning, packed with as-yet unrealized potential, while "fruit" is not a package of potential but a reward for achievement—doesn't each of us wish to enjoy the *fruits* of our labor?

There is a simple reason why fruit is a metaphor for reward: evolution designed fruit to appeal to us animals that way. The metaphorical meaning of "fruit" has its origins in the primate predilection for these sweet, nutritious gifts of nature. But it turns out that we may owe a great deal more to the primate love of fruit than just the odd metaphor. Our evolutionary ancestors' diet of fruit may also be responsible for primate color vision. Before we explore such anthropocentric aspects of fruit, we ought first to consider why fruit evolved. From the plant point of view, what is fruit for?

Strictly speaking, this is the wrong way to pose an evolutionary question because to ask what something is *for* implies purposefulness on the part of natural selection. Natural selection has no purpose. It is a blind mechanism that favors any inherited trait that successfully transmits itself to future generations. However, even professional evolutionary biologists lapse into asking what things are for. It is a convenient shorthand for more long-winded but accurate questions such as "How does the possession of fruit increase the genetic contribution of its bearers to future generations?"

Fruits induce animals to visit the plant that produces them. Seeds may then be carried away, in the jaws or paws of an animal if the fruit is large, or in its gut if they are small bird-fruits such as berries. Seed dispersal is so obviously "what fruit is for" that the evolutionary question "What is the advantage of dispersal?" may not have occurred to you. The answer to this deeper question is not as obvious as it might seem. The question is also a very general one, because all organisms, not just plants and certainly not just fruit trees, have mechanisms of dispersal. Dispersal must confer a very powerful benefit for it to be universal among living things.

At first sight there are at least two very good reasons for *not* dispersing your offspring. First, your own successful reproduction as a parent proves that the home site is a good place for individuals like you. If you did well there, so should your offspring. Other places, not tried or tested, carry no guarantee of success. Second, dispersal itself is very dangerous for juveniles, and the vast majority of them perish in the attempt, either in the jaws of a dispersal agent or a predator (sometimes the same animal), or at the establishment phase when they are deposited somewhere unsuitable. These seem like two very powerful arguments against dispersal. Where, then, lie its advantages?

The answer as to why dispersal is so universally advantageous was discovered in 1977 by evolutionary biologists Bill Hamilton and Bob May using a mathematical model of the process. By representing real-life situations in the abstract language of mathematics you can boil a question down to its absolute essentials, which makes it easier to interpret and generalize the answers the model gives. Hamilton and May's model of dispersal envisages a simple world in which organisms occupy a limited number of habitable sites. To get established, juveniles have to find a vacant site. At the end of a year all adults die, vacating their sites, and their offspring are then released. Hamilton and May imagined that there were two kinds of organisms with different dispersal behaviors. In their honor I'm going to call one "Bill" and the other "Bob" (but ignore their gender please—plants tend to be hermaphrodites). You can think of a model like this as a game in which Bill plays Bob to see who can win the most sites for his offspring.

Bill scatters a fraction of his offspring around, while Bob is a homebody who keeps all his seeds in his own site. There are always more juveniles than establishment sites, so juvenile mortality is severe. Which type of organism would you back as the winner in competition for the most sites, the disperser Bill or the nondisperser Bob? Hamilton and May's model showed that the disperser always won, even when dispersing juveniles suffered very severe mortality. The reason is that both dispersers and stay-at-homes had a high probability of recolonizing their home sites, but that only dispersers had any chance of colonizing new sites. Even when this chance was very, very small, it gave dispersers the edge over stay-at-homes because the latter could never increase their share of the territory. Obvious, once you think of it, isn't it?

In the real world, there are additional hazards of staying near home, and there are also ways in which the hazards of dispersal can be reduced. Both of these favor Bill's strategy over homebody Bob's. An important hazard of staying home is that mother trees are the headquarters for specialized natural enemies like caterpillars and fungal diseases that have a particular taste for the species. Dispersal is a way of escaping these specialized enemies. I discuss the far-reaching consequences of this for plant biodiversity in my book *Demons in Eden*. The hazards of dispersal can be reduced if a tree can direct its seeds to favorable sites for establishment by attracting particular dispersal agents that can carry seeds there. Until recently, directed dispersal was believed to be beyond the power of plants, but more and more cases are now coming to light. Mistletoes are the best known example.

Mistletoes (of which there are many species) are parasitic on the branches of trees, so their seeds must somehow be deposited on this aerial habitat if they are to germinate and establish successfully. Mistletoes, like the European species *Viscum album*, solve this problem by producing berries containing an extremely sticky jelly that can survive passage through the gut of a bird. When a bird that has fed on mistletoe wipes its beak on a branch or defecates, seeds become glued to the branch, where they will later germinate. The birds that feed on mistletoe seem to be unusually faithful to their food plant, and several are named after it. The mistle thrush *Turdus viscivorus* especially favors *Viscum album*, and in parts of Europe where the plant is common a mistle thrush will stoutly defend a tree containing mistletoe from other birds during the berrying season.

In Australia the mistletoe bird *Dicaeum hirundinaceum*, which feeds on the mistletoe *Amyema quandang*, perches on branches of just the right diameter for the establishment of mistletoe seedlings. Thinner branches are unable to survive parasitism by mistletoe, while larger ones have a bark that is too thick for seedlings to penetrate with their parasitic rootlets. Other birds like jays and nutcrackers that disperse the seeds of certain pines, and even ants that disperse the seeds of smaller plants, provide further examples of directed seed dispersal. The behavior of all these animals raises the odds that a dispersing seed will land in a propitious location. Like the realtor said, only three things matter: location, location, location.

Now that we understand the imperative to disperse, it is easier to comprehend the evolution of fleshy fruit. The story begins a long way back, before the origin of either flowering plants or mammals. The first fleshy upholstery around seeds was not technically a fruit (this term is only used for flowering plants) such as Andrew Marvell's apples, grapevines, melons, or peaches, but belonged to early seed plants like ginkgo and cycads. The very distinctive, fan-shaped leaves of ginkgos have been found in fossil deposits more than 250 million years old. Because the squishy tissues of the fruit did not fossilize, we cannot be sure that these early ginkgos possessed the fleshy, fetid coverings of their modern descendants, although flesh would no doubt have attracted late Paleozoic reptiles. Later, during the Mesozoic era, between 250 and 265 million years ago, reptiles, including dinosaurs, would also have been the main consumers of flesh-covered seeds. But how good was their color vision? Could they have seen the blush of a berry or the gleaming skin of a ripe fruit? If the color vision of birds, modern descendants of dinosaurs, is anything to go by, the answer is that their color perception was a lot better than ours.

Color, like beauty, is in the brain of the beholder. The colors we and other animals perceive in our environment are not "out there," but are constructed in our brain from sensory inputs. Understand those sensory inputs, and it's possible to compare the visual capabilities of different individuals and different species. Birds and reptiles can see light with wavelengths from the ultraviolet (310 nanometers: a nanometer is a billionth of a meter) to red (700 nanometers). Vision in mammals, including ourselves, is confined to the range 400–700 nanometers, so birds and reptiles can see ultraviolet wavelengths (310–400nm) that are invisible to us.

The range of wavelengths that an animal can perceive is determined mainly by the number of specialized types of light-receptors in the retina of the eye. In very dim light, vision depends on cells called rods. These are highly sensitive to light in the middle wavelengths of our range and they produce a monochrome image. We cannot see colors in very dim light because there is only one kind of rod. To perceive color, the brain must receive inputs from at least two different types of receptor, each responding to light of its own characteristic range of wavelengths. The reason we cannot perceive color with inputs from only one receptor type is that a receptor cell responds to a range of wavelengths, and stimulation by light anywhere within this range will send the same signal to the brain, regardless of the wavelength (color) it receives. Only when the brain has separate inputs from cells with different wavelength sensitivities can a comparison be made between them and can color be perceived as a result.

The specialized receptor cells used in color perception are called cones. These operate in brighter light than rods and come in different types. In most birds, there are four types of cones, with peaks of sensitivity in the ultraviolet (UV), blue, green, and red. The characteristic wavelength sensitivity of cones depends upon a light-sensitive protein called an opsin that changes shape when stimulated by light, triggering a nerve impulse to the brain. Very slight but crucial chemical differences between opsin molecules alter the wavelengths at which they trigger a nerve impulse. Each cone type contains a specific opsin.

While most birds, reptiles, and even goldfish have four cone types, giving them four-color vision, the majority of mammals, including for example dogs and horses, have only two cone types. Dogs and horses have no cones that peak in the mid (green) part of the spectrum. Green light stimulates the same cones that are stimulated by red light, so dogs and horses are red-green colorblind. So before you set off on a romantic, horse-drawn carriage ride around Central Park in New York, here's some advice: check that your driver is not red-green color blind (surprisingly common in human males), because if he is, his horse can't tell red-for-stop from green-for-go at traffic lights either. Better find yourself a conveyance in which at least one pair of eyes up front can sense red for danger.

How did mammals, descendants of a vertebrate lineage in which even the lowly goldfish have four-color vision, become debased to only two? And how did primates, which almost uniquely among mammals have three cone types, reacquire the extra color sensor? The answers lie in our evolutionary history. (And yes, it *really does* have something to do with fruit).

The mammal lineage is rooted in the Mesozoic, the age of

the reptiles, when mammals were small, insectivorous shrew-like nocturnal creatures. As nocturnal animals, early mammals would have depended on rods rather than cones for vision, and the loss of two cone types would not have been missed. It is even possible that two-color vision could have been an advantage. In a study that compared human subjects with two-color vision (who were red-green colorblind) with normal subjects, the red-green colorblind subjects were better than people with normal color vision in spotting patterns that were camouflaged by color. Lab studies with nonhuman primates have obtained the same result. The experimental conditions of such studies are a long way from the Mesozoic, but one might imagine a mutant, two-color-vision mammal of that era foraging in the gloaming and finding it easier than primitive animals with four-color vision to spot color-camouflaged insect prey. Such an advantage could lead to the spread of two-color vision by natural selection. *Use it or lose it* is a common theme in evolution, with natural selection often finding better uses for the resources that would otherwise go into redundant organs or cell types.

However it may have happened, mammals did apparently lose two of their cone types during their early evolution (see figure 9.1). We still lack vision in the UV, but as primates we regained cones that are sensitive to green light. Our three-color vision has receptors that peak in the red, green, and blue parts of the spectrum. Any color display on a computer or TV is witness to this fact. Electronic displays can reproduce millions of colors and they produce images that we find true-to-life by combining microscopic pixels of red, green, and blue. If electronic displays were designed specifically for dogs (it can only be a matter of time), they could reproduce images that would look lifelike to a dog with just blue and red pixels, because dogs have only two cone types. Goldfish TV, of course, would need pixels in the red, green, blue, and UV. But I digress.

Back at end of the Mesozoic, 65 million years ago, the world

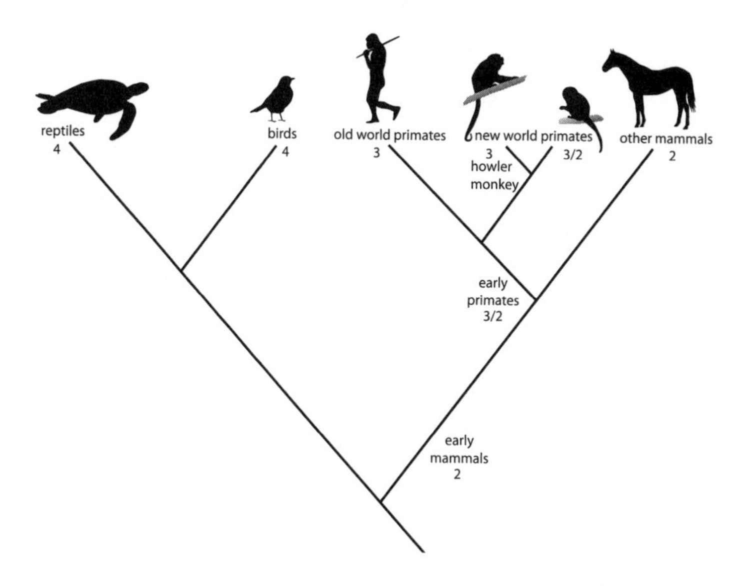

FIGURE 9.1. Evolution of color vision in primates and other vertebrates. Numerals indicate the number of cone types (3/2 indicates polymorphic vision; see text). Lengths of branches in the evolutionary tree are not proportional to time.

of early mammals changed forever when the dinosaurs, which had dominated life on Earth for an entire geological era, went extinct. The demise of the dinosaurs heralded the rise of the mammals. With the dinosaurs gone, the daylight no longer belonged to the reptiles, and diurnal mammals evolved, among them diurnal primates. What did they eat? Fruit makes up a large part of the diet of most living species of monkeys and apes, suggesting that early primate ancestors also depended upon fruit. The fossil teeth of hominids, such as *Australopithecus afarensis*, believed to be an ancestor of our own species, show adaptation for fruit-eating too.

Imagine the period of transition from nocturnal foraging for insects to diurnal foraging for fruit. The majority of fruit available would have coevolved with birds or other dinosaurs with four-color vision. How would a primate with two-color vision manage? We can guess what the fruit would have looked like from what modern birds like to eat. The fruits that birds are attracted to are mainly red or black in color. Bilberries, for example, appear almost black to our eyes, but are UV-reflecting, and birds use this wavelength to hunt for them. A bilberry to a bird is as conspicuous as a red holly berry is to us. Now imagine an early primate, red-green colorblind and unable to see in the UV foraging for small fruit that was adapted for bird dispersal. How successful would they have been, especially in competition with birds?

If you had to redesign the mammalian two-color visual system for a fruit-eating primate, what would you do? Fixing the red-green color blindness so that primates could see red fruit against a background of green leaves is one obvious solution. This is apparently the answer that evolution came up with too. Evolution is a gradual process and the transition from two-color to three-color vision in primates occurred in two distinct steps. The first step involved a mutation in the gene that produced the normal red-sensitive opsin protein. The mutant version of the gene produced an alternative version of the protein that was sensitive to green light.

The alternative versions of a gene are called alleles. Humans and other mammals have two copies of every gene, one on the chromosome inherited from mother and one on the chromosome inherited from father. So if there are two different alleles in a population, an individual might have either two copies of the same allele or one copy of each, depending on what has been inherited from either parent. A population that contains more than one allele is said to be polymorphic for the character that is coded by that gene. Primate populations with red and green alleles have polymorphic color vision. Which colors a particular individual in such a population can see depends on which alleles it has.

It so happens that the gene producing red or green opsins is situated on the X chromosome, which complicates polymorphic color vision in an intriguing way. X is a sex chromosome that in primates and other mammals is present as two copies in females, but only as one copy in males. This means that males can have only one opsin allele (red or green), but females can have one or two (two red, two green, or one of each). So in New World monkeys, where this situation occurs, females may have the allele producing the red-sensitive opsin on one X chromosome and the allele producing the green-sensitive opsin on the other. Combined with the blue receptor (which is not on a sex chromosome), this gives such females three-color vision, because they have cones in their retinas for blue, green, and red light reception. Males, by contrast, always have two-color vision because they have only one X chromosome and therefore can have either the green opsin or the red, but not both. Females with two copies of the same allele also have two-color vision, of course.

Now we can return to the question of fruit. How much benefit do female monkeys with three cone types get from their extra color sensor? By comparing the known spectral responses of the different opsins with the colors of fruit and the leafy backgrounds against which fruit are displayed, it is possible to get a very good idea of what monkeys with two cone types or three cone types can actually see. Studies of this kind have been verified by behavioral experiments, but otherwise need not involve direct experiments on animals. The results suggest that there are two advantages of three-color vision: red fruit show up much better against green leaves, especially in dim light, and fruit ripeness, which is signaled by the reddening hue of fruits, is easier to judge. Even humans with defective color vision have difficulty finding fruit among foliage.

You will recall that I mentioned at the start of this discussion of three-color vision in primates that there were two stages in its evolution. Visual polymorphism is only the first stage and it is clearly incomplete; perhaps, as a male, I can even venture the opinion that it is imperfect, since no males and fewer than half of any polymorphic population have three-color vision. However, among all the New World primates, one species, the howler monkey, has evolved full three-color vision.

In howler monkeys the opsin gene on the X chromosome has been duplicated, so there are two on the X rather than just the one that other New World monkeys have. One of the opsin genes on the X produces the green receptor and the other produces the red. Because both these genes are on the X, even male howlers, that of course have only one X chromosome, have three-color vision. Comparison of the howler monkey visual system with that of a related polymorphic species suggests that in the case of the howlers, the greatest benefit they get from three-color vision is an increased ability to find young, red leaves that form an important part of their diet, rather than an increased ability to find fruit.

How did full three-color vision evolve? This is an interesting question because it actually happened twice, once in New World howler monkeys and once much earlier in an ancestor of all the Old World primates, including ourselves (see figure 9.1). All Old World primates have full three-color vision. In howlers, the duplication of the opsin gene on the X chromosome appears to have happened through an error during the normal process of recombination that takes place between chromosome pairs during the production of eggs (and sperm). Normally in recombination, equal lengths of chromosome are swapped, but if the swapped pieces are unequal in length, one chromosome will end up with extra genes and the other with a deficiency. Analysis of DNA sequences shows that that is what must have happened between two X chromosomes (and therefore in a female) in a common ancestor of all the howlers, duplicating the opsin gene. Presumably the two opsin genes on the howler's X chromosome produce different color receptors, because one of the original recombining Xs carried a red allele and the other a green one. The advantage of possessing both these sensor types on a single chromosome would explain how a rare genetic event like this could establish and spread to become a fixed characteristic of the entire species.

Diet seems to have played a role in the evolution of three-color vision in both primate groups in which it occurred. In howler monkeys it appears to have been driven by their taste for young, red leaves. In Old World primates it was their diet of fruit. The ancestor of all the Old World primates from which we have inherited three-color vision also duplicated the opsin gene on its X chromosome. However, the duplication of the opsin gene in this case seems to have involved some natural genetic engineering by a transposon that is able to skip around the genome, duplicating itself. Our chromosomes are littered with such DNA parasites and one of them, called Alu, is found in suspicious association with the opsin gene, implicating it in the unequal exchange between chromosomes that created an extra copy on the X. This must have happened about 40 million years ago, before the evolution of all the species of Old World primate that are alive today, because all have three-color vision like us. If we could go back in time to find the common ancestor of both Old World and New World primates, we would probably find that it had a polymorphic visual system of the kind that has persisted in most New World monkeys.

Color is not the only signal of ripeness in fruit, but its effect on the human psyche is not to be underestimated. Supermarkets know that they can sell bright red, unripe tomatoes and peaches that have been bred to lie to the consumer about their ripeness, because the color makes them look good to us. The sign over the fruit display that says "Do not squeeze me till I'm yours" may sound cute, but it is also a recognition that we don't trust our eyes when it comes to what the supermarket has to offer. We have an urge to feel the fruit, sniff it, and taste it too, just like any other primate. Nonetheless, we can justifiably claim that thanks to our evolutionary history, fruit truly are a feast for the eyes, as well as for the stomach.

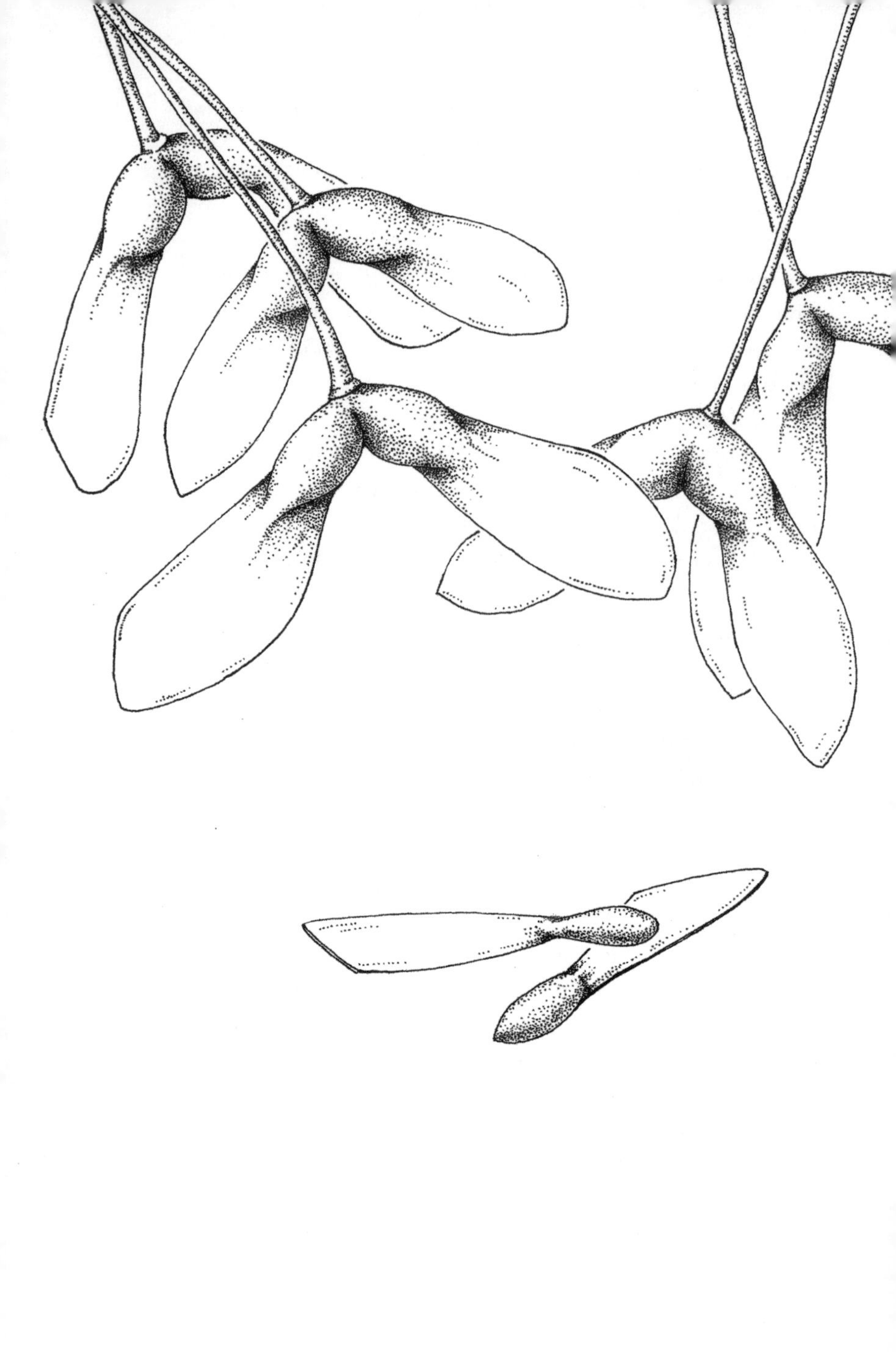

{ 10 }

Winged Seeds

DISPERSAL

O wild West Wind, thou breath of autumn's being . . .
Who chariotest to their dark wintry bed
The winged seeds, where they lie cold and low,
Each like a corpse within its grave, until
Thine azure sister of the spring shall blow
PERCY BYSSHE SHELLEY, FROM "ODE TO THE WEST WIND"

For most of human history people have surely gazed skyward at birds soaring in the blue firmament and dreamed of flying. But nature inspired us to an ambition that she had not equipped us to fulfill. The flying beings we conjured from our imagination were fancifully endowed with feathered wings, as though possessing these magical appendages were all that was needed to let the Sphinx of ancient Egypt fly, to carry Mercury aloft by his winged sandals, to levitate the demonic mounts ridden by the Valkyries of German legend, or to take angels to heaven. Bird flight motivated the human desire to fly, but flapping like a bird could never supply the necessary motive power. When we finally conquered the air, it was not by imitating birds but by learning to parachute like thistledown, to glide like a winged seed or to helicopter like a maple fruit. We aspired to be birds, but have flown by emulating plants.

It's unlikely that when Orville and Wilbur Wright were growing up they told their mother, who was the engineer of the family, that it was their ambition to fly like a seed, but that's how they made it into the air. The fixed wings of their wooden craft were clothed in a fine cotton fabric made from the very fibers that cotton seeds use to fly. Of course, plants evolved flight long before the Wright brothers gave up cycle repair for higher ambitions, and maple seeds traveled by helicopter well before Leonardo da Vinci sketched such a vehicle or his ancestors could even claim to be *Homo sapiens*. Indeed, conifer seeds have flown since before birds evolved from their earthbound ancestors, the dinosaurs.

The supreme flyers among seeds are the gliders made by the tropical vine *Alsomitra macrocarpa* that lives in the tropical forests of Southeast Asia. The plant belongs to the squash family and climbs five meters into the rainforest canopy, from where hang fruit the size and shape of large footballs. When ripe, the fruit splits open and as its contents dry, the winged seeds drop from beneath and glide on filmy, translucent wings that are twelve centimeters from wingtip to wingtip (figure 10.1 c). The pea-sized seed in the cockpit of its glider may be carried hundreds of meters, some even reaching the decks of ships at sea if there is a favorable wind. Favorable to dispersal, that is, but not necessarily to the seed's future. Too distant dispersal can be a bad thing.

The shape of the *Alsomitra* seed inspired the Austrian Igo Etrich to design a glider with wings of the same shape (figure 10.1 a). The flying-wing design, lacking a tail, is highly stable, as one early English aviator tried to demonstrate by flying hands-free, nonchalantly ignoring the plane's controls to scribble a note while in flight, and nearly crash landing as a result. The world record for the longest time aloft for a model paper airplane (over eighteen

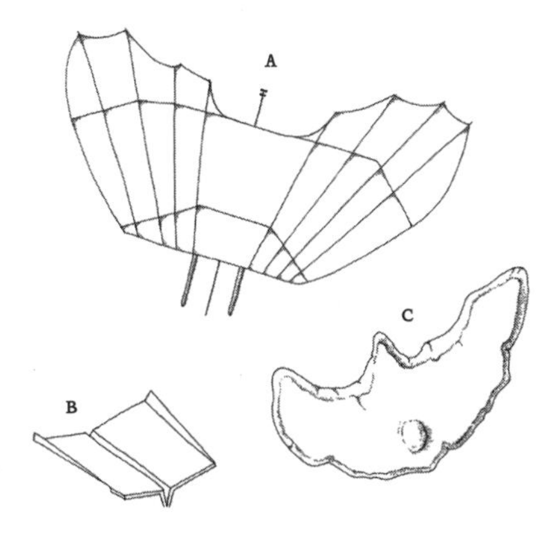

FIGURE 10.1. (A) The wing pattern of an aircraft wing inspired by the seed of *Alsomitra macrocarpa*, (B) the flying-wing design of the world's best paper airplane, (C) A seed of *Alsomitra macrocarpa*.

seconds) is held by a tailless flying wing (figure 10.1 B), illustrating the aerodynamic stability of the form evolved by *Alsomitra*'s glider.

Unlike paper airplanes or gliding seeds (but like birds) planes need to be maneuverable. This is a disadvantage for a flying wing that is so stable that it can fly itself. Etrich solved the problem by adding a birdlike tail to the *Alsomitra*-inspired wings, producing a powered, maneuverable aircraft design that was so successful that it was mass-produced for military use in World War I. Gliding is in fact rare among seeds, possibly because it works best as a dispersal strategy in descending flight through still air. These conditions are unusual. Significantly, the other notable example of gliding seeds occurs, like *Alsomitra*, in a vine of tropical forest habitats where the cavernous space beneath the canopy gives flying room and is protected from strong winds.

Seeds that fly on a single, unilateral wing are much more common than those that glide. The fruits of ash (*Fraxinus* spp.) hang from the boughs like bunches of keys before they fly; maple (*Acer* spp.) fruits are born as conjoined twins that separate before takeoff. All except the plumpest pine seeds have a wing. (The fat, wingless ones are dispersed by animals). Unilateral wings have evolved many times in different plant groups. The bigger the seed, the larger the wing that is required to transport it. The ratio of the mass of the whole seed divided by the area of the wing is called the wing loading. The record for wingspan is held by the Brazilian zebra wood tree (*Centrolobium robustum*) whose large seeds are protected by spines and furnished with a wing up to thirty centimeters long. You would definitely want to duck if you saw one of those spinning toward you through the forest!

So far as I know, flight on a unilateral wing has never been deliberately tried by human aviators. They probably wouldn't enjoy the experience of being spun like a top on one wing and a prayer. Helicopters require special design features to prevent the aircraft spinning in the direction opposite to the rotor. The advantage of this type of spinning flight for seed dispersal is that as well as slowing their descent in still air it can carry seeds upward when the wind blows.

Though so obviously equipped for flight, many dandelion, willow herb, or cottonwood seeds do not seem to travel far. Witness the ground around such plants that is typically littered with the downy residue of failed seed dispersal. A lot of parachute seeds are evidently not very good fliers, and the same goes for winged seeds. This evidence is misleading, however, because the evolution of seeds is governed by the laws that apply to large numbers, not by the fate of the average seed. Just as it is the exceptional seed, the one-in-a-million that escapes the jaws of animals, which establishes the next generation, so it is the flying ace and not the prematurely grounded dunces of flight that carries the genes of its parents into the future.

What makes a flying ace? Chance is certainly part of the answer, which is why so many seeds are made: many must be launched for any to succeed. A low wing loading also helps, but parental behavior plays a role too. Studies of seed dispersal suggest that dandelions and other plants do not simply release their seeds at random. Wind-dispersed seeds remain attached to the plant until a wind of sufficient strength is able to whisk them away. In trees, the springiness of branches that bounce in strong winds may also help launch seeds at the right moment for them to be carried upward by air turbulence. Animals have learnt this lesson too. Orangutans use their weight to deliberately set branches swaying so that they may leap large gaps between trees with less effort. In gymnastics, pole-vaulters use the same trick to propel themselves skyward. Seeds, orangs, and gymnasts all use the springiness of wood to aid flight.

Surprisingly, the importance of gusts of wind and air turbulence to the long-distance dispersal of wind-dispersed seeds is a recent discovery. We owe it to a biologist of rare imagination. Henry Horn, a professor at Princeton University, has for decades looked at the natural world with an idiosyncratic gaze, and even at one time through spectacles that mimicked butterfly vision. When I visited him in the mid-1990s a label on his door read "Boy-Wonder Emeritus," a self-deprecating reference to a pathbreaking book he wrote when still a graduate student called *The Adaptive Geometry of Trees*. At the time I visited, Henry was experimenting with a simple, homemade apparatus that enabled him to study winged seeds hovering in stationary flight. His idea was that winged seeds could be carried long distances on wind eddies, like surfers riding a breaking wave. Models showed it could work, but skeptics questioned whether wind eddies were prolonged enough to transport seeds very far.

Henry Horn solved this problem in typically ingenious fashion by making field observations in Central Park, New York. In 2005 the artists Christo and Jeanne-Claude created an installation there, where they erected 7,500 gates placed at four-meter intervals along pedestrian footpaths. Each gate formed an archway about five meters high from which hung a loose curtain of saffron-colored fabric that was free to flap in the breeze. Quite fortuitously, the art installation provided a perfect field laboratory for studying the duration and path of wind eddies. By photographing a stretch of gates, Horn discovered that some eddies could last for more than a minute and a half, sweeping the curtains along a hundred meters of path. His calculations showed him that such eddies could transport seeds half a kilometer from their natal tree, much farther than was previously thought.

It is one thing for a seed to be transported a long distance, but quite another for the traveler to successfully settle down and raise a family. How do we know that long-distance seed transport is successful? The evidence, strangely enough, comes not from the present but from the past. If you live in the temperate zone, which includes the bulk of North America and most of Europe, the wild woods where you probably like to hike or walk your dog hide a secret. The trees may look old and venerable, and they may indeed be so measured by the yardstick of human longevity, but measured in tree-time the species are newcomers.

Six lifetimes of the redwood General Sherman ago, or about twelve thousand years before the present, the temperate zone was covered by ice and treeless tundra. The geological signs of the last ice age are there to be read in landscapes shaped by glaciers, ice sheets, and torrential meltwater, but there are botanical telltales too. As the ice retreated northward, trees followed in its wake, and we have their footprints cast in buried pollen grains to show how rapidly they advanced. These footprints tell the same story for nearly all forest trees: trees recolonized the temperate zone much more quickly than the slow, creeping progress that you would expect from the average dispersal distances of seeds. The pace of tree migration was not set by the average seed traveler, but by a vanguard of the farthest of the far-flung. They leaped, they survived, and they grew.

In species like oak, seeds would have been carried to the frontier and beyond by birds such as jays. Winged seeds like birch, ash, and maple must have surfed the wind to new terrain. So recent in tree-time is the recolonization of the temperate zone that some species are still moving northward, and the ongoing dispersal process has left a detectable evolutionary legacy in its wake.

Lodgepole pine, that favorite snack of red crossbills, spans the length of the West coast of North America, from Baja California in the South to the Yukon in Canada. In the frozen North, lodgepole pine is captured in the act of colonization. At the northern limit of the species' range, populations may have arrived only a century ago. One of these new northern populations, consisting of only forty-five trees, is seventy kilometers from the nearest population to its south, representing the latest in a series of long-distance leaps by which lodgepole pine must have progressed up the West Coast. As it migrated north by leaps and bounds, the weight and the wing loading of seeds progressively decreased. Seeds in the recent populations of the Yukon have a wing loading that is about 25 percent less than those of Baja.

Thus, long-distance dispersal repeatedly winnowed the seed of lodgepole pine, with each successive population being founded by seeds that were lighter and flew better than their ancestors.

Just as natural selection favors better-dispersed seeds when long-distance colonists can found new populations, it favors reduced dispersal when the life chances of long-distance travelers are nil. Thus paradoxically, although plants require good dispersal powers to reach islands, once they have arrived there and become established natural selection often takes away the very dispersal powers that carried them there in the first place. The islands of Hawaii and other remote archipelagos are typically populated with species belonging to the dandelion family that have lost their parachutes, hooks, and other paraphernalia that might carry them away and land them in the drink. As for those seeds that make a more propitious landing, Shelley's winged seeds that lie cold and low, "each like a corpse within its grave," they now face another challenge before they see the sun.

{ 11 }

Circumstance Unknown

FATE

Longing is like the Seed
That wrestles in the Ground,
Believing if it intercede
It shall at length be found.
The Hour, and the Clime–
Each Circumstance unknown,
What Constancy must be achieved
Before it see the Sun!
EMILY DICKINSON, "LONGING IS LIKE THE SEED"

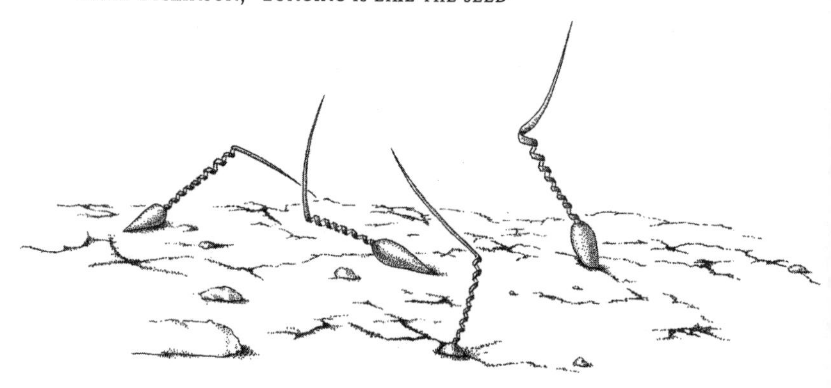

The tragedy of parenthood is that parting with offspring is necessary to their future. They must strive for themselves. Mercifully, we are not usually forced to cast them adrift during childhood, but it was not always thus. In eighteenth-century London a thousand babies a year were abandoned on the streets to die because their mothers could not support them. A mother's only desperate hope was that her baby might be one of the lucky few to be taken in at Thomas Coram's Foundling Hospital. Parting might mean survival for the child and a possible reunion in better times to come. One Coram mother of the 1750s left her baby with this poem in explanation of her misfortune and hopes:

Hard is my lot in deep Distress
To have no help where Most should find
Sure Nature meant her sacred Laws
Should men as strong as women bind
Regardless he, Unable I,
To keep this image of my heart
'Tis Vile to Murder! hard to Starve
And Death Almost to me to part
If fortune should her favours give
That I in Better plight may live
I'd try to have my boy again
And Train him up the best of Men.

The site of the Foundling Hospital now houses a museum, and among the paintings given to the institution by fashionable artists and patrons of Georgian London is a small display case containing tokens that mothers left with their babies. To form a bond, however slender with her child, one mother left a brass button, another a key, a third a hairpin, a tiny fish carved from bone, a crushed and unusable thimble, an enamel decanter label inscribed "ALE." Most poignant of all the tokens of fractured motherhood, testament to the grinding poverty of its giver, is a single, empty hazelnut shell with a small hole gnawed in it by a mouse. This seed was faithfully preserved for the link it might one day have forged between parent and offspring.

I visited the Foundling Museum on a bright spring day, but left in somber reverie. George Frideric Handel, a stalwart patron of the Foundling Hospital, used to give annual charity concerts of his oratorio *The Messiah* there, but no hallelujahs entered my thoughts that afternoon. Instead, I contemplated the millions of plumed seeds, piling in tawny drifts upon the ground, products of the huge plane trees with which the city fathers of Georgian London endowed their elegant squares. The London plane was also a foundling, the hybrid offspring of an unrecorded mating between American and oriental species of *Platanus*. It still abandons its offspring by the billion on the streets of London.

The uncertain fate of seeds has been used in literature as an image of the vagaries of human existence. In his novel *The Grapes of Wrath*, about the human tragedy created by the ecological disaster of the Great Drought in the 1930s, John Steinbeck devotes the whole of the third chapter to a minute description of how the seeds of plants by the side of the highway are dispersed. The highway is the one that the Joads, a family of impoverished Oklahoma sharecroppers, will travel to flee the dust and debt of their home state for the promised land of California.

The concrete highway was edged with a mat of tangled, dry grass, and the grass heads were heavy with oat beards to catch on a dog's coat, and foxtails to tangle in a horse's fetlocks, and clover burrs to fasten in sheep's wool; sleeping life waiting to be spread and dispersed, every seed armed with an appliance of dispersal, twisting darts and parachutes for the wind, little spears and balls of tiny thorns, and all waiting for animals and for the wind, for a man's trouser cuff or the hem of a woman's skirt, all passive but armed with appliances of activity, still, but each possessed of the anlage of movement.

In the novel, a land turtle appears and as it drags itself over the ground some seeds become lodged inside its shell. A passing truck driver swerves to try to hit the turtle, but just clips its shell, and the animal is flipped into the air like a tiddlywink, and it spins off the road, where it lands on its back. With great effort, the turtle rights itself, and as it flops onto the ground, a wild oat seed falls out and three spearhead seeds stick into the earth. Finally, the seeds are buried as the turtle plows over them, covering them in soil. The same denuded soil that the sharecroppers can no longer rely on to support them. As the story unfolds in subsequent chapters, the Joads also have a tough time on this highway. Not all of them make it to California, and those who get there find it is not the paradise they expected.

For refugee or seed, dispersal is a haphazard and hazardous business. But soil is a refuge for dispersed seeds and many are equipped to bury themselves or to induce animals to do this for them. Seeds that are tiny enough slip down cracks in the soil surface and are buried simply by gravity. Some larger seeds, especially those of grasses, have bristle-like appendages called awns that cause them to fall like a dart onto the soil surface, where some will chance to enter upright into a crevice. My favorite among these grasses is the very common wall barley *Hordeum murinum*, whose seed heads are darts with perfect flights fashioned from numerous long awns. The awns have minute barbs on them, so the whole seed head will not only fly perfectly toward its target when you launch it at a nearby relative, but it will cling to his clothing too. Not for nothing is this particular grass closely associated with human habitat.

Wheat and barley also have long awns, though crop breeding has rendered them redundant because cereals are bred to prevent the seed heads from breaking up and scattering their seeds before the harvester can collect them. In agriculture, cereal grains are sown naked, stripped of their awns and the remnant flower parts that are the normal apparel of grass seeds when dispersed in the wild. The wild ancestor of barley (*Hordeum spontaneum*) has a long, stiff awn which vibrates continuously in the wind, pushing the erect seed more firmly into its crevice like a pile driver. But there is a limitation to this device: a seed must first find a crevice. An awn can come in useful here too. In some wild grasses the awn is hygroscopic, which means that it takes up moisture from the air and when it does so it changes shape (human hair has the same property). As its hygroscopic awn twists and untwists with changing humidity, the seed is propelled across the soil surface. It travels like a wayward gondola steered by a deranged boatman wielding a spasmodically contorted punting pole until the seed falls or is pushed into a crack. Even then the awn may not have finished its work. A few species have tightly coiled hygroscopic awns furnished with barbs that give them purchase on the sides of a crevice, and these are able to drill themselves into loose soil. Who would have thought that a simple bristle on a grass seed could pilot it to a vertical landing, punt it over the soil surface, and drill it into the earth?

What a plant cannot do for itself, it can often trick or bribe an animal into doing for it. Ants are the favored agents for burying seeds. Seeds of violets, primulas, and a great assortment of over three thousand species belonging to more than eighty different plant families have a fatty wart called an elaiosome attached to them that attracts ants in search of food. Patrolling ants that find such seeds do not detach the elaiosome there and then, but carry the whole seed back to their nest, where it is buried. Once the seed has been stripped of its elaiosome in the ant nest, the ants dump it in a viable state on a trash pile where it can germinate. Seeds thus buried are hidden from predators such as birds and have a greater chance of survival than seeds that have not been transported by ants.

In Australia, where elaiosomes are so common they are almost de rigueur in parts of the native flora, stick insects have evolved eggs that mimic the appearance of seeds. They even have a fat-rich structure like an elaiosome that causes ants to carry them away to the protection of their nests, where they are hidden from parasites. What are eggs, after all, but a kind of seed? The chemical composition of elaiosomes is interesting because it is quite different from that of fruit and seems to be tailored to the food preferences of ants, whose normal prey are insects. The importance of ants in seed burial, as distinct from dispersal, is emphasized by the fact that many tropical trees and some Mediterranean shrubs have fleshy fruits that contain seeds furnished with an elaiosome. The whole is like a package made for a game of pass-the-parcel, with a gift for birds on the outside layer and another for ants concealed inside. Birds consume the fruit and void the elaiosome-bearing seeds. Dispersed seeds are then collected from bird feces and buried by ants.

If fleshy fruits, wings, hooks, and barbs are the instruments by which seeds disperse themselves in space, then dormancy is how seeds disperse themselves through time. The ground beneath our feet and the soil beneath the plow are replete with seeds of species that have accumulated there over tens, even hundreds of years. As Henry David Thoreau described it, "The very earth itself is a granary and a seminary so that to some minds, its surface is regarded as the cuticle of one living creature."

Dormant seeds are time travelers. They are able to shut down their metabolism and to tick over in a state of quiescence, alive but consuming almost no energy, for years, decades, and, in a few cases, even centuries. Contrary to popular belief, the oldest seeds ever germinated were not wheat seeds buried in ancient Egypt with the pharaohs. These seeds have never been successfully germinated. Current holder of the title "Oldest seed ever to germinate" is a two-thousand-year-old date seed retrieved from archaeological excavations of King Herod's palace at Masada in Israel, where Jewish Zealots held out against a Roman seige, finally taking their own lives rather than surrender. Seeds found during excavations lay in a drawer for thirty years until someone thought they might try to germinate them, little thinking any would be viable. To everyone's astonishment one was.

Time takes its toll on DNA, unless it is deep-frozen. For this reason the Millennium Seed Bank at the Royal Botanic Gardens, Kew, keeps its seed samples in freezers at minus twenty degrees Celsius inside a bomb-proof vault. If England gets too warm for comfort (or if someone forgets to pay the electricity bill), the Norwegian government has built a doomsday vault for samples of the world's crop seeds that are to be kept in a naturally frozen chamber 120 meters inside a mountain on the island of Spitsbergen, inside the arctic circle.

Seeds of plants in the pea (Fabaceae) are naturally blessed with a highly resistant seed coat and are often long-lived. Thus it was that in 1940 when the Natural History Museum in London was bombed and the fire brigade played their hoses upon the ashes, seeds of the legume *Albizia* cheerfully woke up and germinated on the herbarium sheet where they had been placed in 1793. More recently, the seed bank at Kew found some seeds among documents belonging to a Dutch merchant who had visited South Africa two hundred years ago. On the way back to Europe the merchant's ship was captured by British pirates and his papers ended up in the National Archives in London. Seeds of thirty-three species collected by the merchant were tested, and of those, three species were germinated, two species of legume and a member of the family Proteaceae .

Gardeners know that some vegetable seeds will keep for a couple of years or more, while others have to be bought afresh each season. Lawrence D. Hills, the late English doyen of organic horticulture, even wrote a poem on the subject in the style of a sixteenth-century poet who wrote advice for farmers in rhyme:

You have in your drawer since Christmas Day,
All the seed packets you daren't throw away.
Seed Catalogues cometh as year it doth end.

But look in ye drawer before money you spend,
Throw out ye Parsnip, 'tis no good next year.
And Scorzonera if there's any there,
For these have a life that is gone with ye wynde.
Unlike all seeds of ye cabbagy kinde,
Broccoli, cauliflower, sprouts, cabbage and kale,
Live long like a farmer who knoweth good ale.
Three years for certain maybe five or four.
To sow in their seasons they stay in ye drawer.

. . . .

Then fillen ye form that your seedsmen doth send.
For novelties plentie, there's money to spend.
Good seed and good horses are worth the expense,
So pay them your dollars as I paid my cents.

The extremes of seed dormancy show what is possible, but not what is normal. It is possible for seeds to survive for very long periods of time, but what is the advantage in doing so? The store of seeds in the soil, which may reach tens of thousands per square meter in cultivated ground (didn't you know it if you have a garden full of weeds!), is often referred to by ecologists as a "seed bank," but this isn't the kind of bank anyone would want to invest their money in. It pays no interest, so that while other seeds germinate, have offspring, and spread their genes, there is no evolutionary percentage at all for sitting dormant in the bank. This is the equivalent of having the value of your money in the bank severely eroded by inflation. Worse, this is a bank regularly robbed by rodents and riddled with fungi that are ready to rot any seed that becomes damaged. Earthworms bury seeds deeper and deeper with time, till all prospect of successful germination is as remote as the soil surface itself. Given the hazards of the soil seed bank, why waste time in such a risky place?

There is really only one possible answer to this question, and the clue to it lies in the fact that dormancy is time travel. The answer is that there are good times and bad times to germinate. A seed only ever gets one shot at germination, so it must get the timing right. The good and bad times vary through the year with the seasons, so the advantage of some short-term dormancy, say between seed set in the fall and seed germination in spring, is not difficult to understand. What is really puzzling is that not all seeds germinate together in spring. Why do so many wait it out in the seed bank for another year? This is where we must think longer-term.

There are good and bad years for germination and growth, and, as Emily Dickinson has it in her poem reproduced at the beginning of this chapter, the hour and the clime are circumstances unknown to a buried seed. If I germinate now, is it going to be a good year or a bad one? What should a seed do in the face of such uncertainty? Interestingly, the answer is different, depending upon whether we ask the seed or the mother plant that produced it. I fancifully say we will "ask" the seed, but of course what I really mean is that we are going to interrogate natural selection on this subject. What does evolution tell us a seed should do? You can probably guess the answer, or you'll kick yourself (or me) when I tell you how simple it is.

A seed should germinate when all the cues available, like temperature and soil moisture, point to conditions being best for growth. Assuming seeds cannot tell what other seeds are going to do and all receive the same cues, then all seeds of a species ought to act the same way and choose the same time to germinate. (If seeds can tell that others are going to germinate, another possible strategy is to wait till they are out of the way so as to avoid the competition, but we'll ignore this complication). So, from the point of view of individual seeds, what is good for one should be good for all. Germination should be synchronous and seed banks should only be transiently populated or empty. In actuality, this is far from the case and seeds in the soil number in the hundreds to thousands per square meter in forests and grasslands, and in the tens of thousands in disturbed habitats like arable fields.

Now let's ask mother. When should your seeds germinate? Mom says, "I have lots of seeds, so I'm not going to put them all in the same basket. I'll have a bit of germination this year, a bit more in two years, some in three, oh! and I'll keep a teensy weensy bit back for five years' time just in case I have a run of bad luck." Governments store grain against emergencies for precisely the same reason. Put the answers of mom and seed side by side and we have a curiously familiar-sounding story of conflict between the desires of parents and the will of offspring. Who wins? You might think it would be the seed, because after all they are no longer attached to the mother plant, so what can she do about it? But plants are cleverer than that!

The mother plant is able to exercise control over her offspring, even after they have flown the coop, because each seed is dispersed in swaddling clothes of maternal tissue. Until it is released from the bondage of the seed coat, the embryo is in thrall to its mother. Mom can program the layers she has wrapped around each seed to be tough or tender, to germinate soon or later, and she often goes for a mixture within her brood. That's why the seed banks are full of grounded offspring, wrestling but unable to escape their maternal bonds till the hour and the clime she has chosen. Many seeds are held there, hostage to an evolutionary insurance policy which says that if you let all your seeds germinate together, one year there will be a disaster and all your offspring will die. It's not that plants are prescient about this or have listened to the sales patter of a life insurance salesperson. Evolution doesn't work like that. Rather, the individual plants that don't spread the risk of total extinction by keeping some of their seeds dormant have been weeded out by natural selection. The risk of total loss of offspring is greatest for annual plants that live only for a year, which is why annuals and other short-lived plants are the most abundant in soil seed banks and the source of the gardener's lament: "One year's seeding is seven years' weeding."

Such is the force exerted by a germinating seed that wooden sailing ships have been rent asunder when their cargos of rice have become wet and started to sprout. Grain silos have been known to split under the pressure of sprouting cereals. Who would dream that such forces slumber in seeds which, if dried to a water content of 5 percent or 10 percent, can be stored for long periods without showing any sign that there is life within?

Germination begins when a seed imbibes water, which initially causes swelling and then embryo growth. Water uptake by seeds is not passive, like filling a bath, but uses the affinity of carbohydrates (e.g., starch) and proteins (e.g., gluten) for water molecules to suck water in like a sponge. The process of water uptake can be remarkably rapid. The quickest seed on the germination draw is the South African klapperbossie *Blepharis mitrata*. Within seconds of contact with water the seed's slick coat of thick, white hairs turns into a slimy, sticky bath mat of swollen, erect fibers that glue the seed to the soil surface. This prevents the seed being washed away by flash floods and also deters ants that would otherwise eat it. Six hours later the embryo has grown sufficiently to rupture its seed coat and has thrust a root into the ground.

Seeds of the herbs *Plantago psyllium* and *P. ovata* have a mucilaginous gum in their seed coats that is so effective at taking up and storing water that the seeds are used in commercial remedies for constipation such as Metamucil and Hydrocil. Chía (*Salvia hispanica*) is an annual herb found in Mexico and the deserts of the American Southwest whose tiny, bean-like seeds also produce a sticky mucilage when they become wet. The seeds were traditionally soaked in water and consumed as a porridge, or ground and made into a seed preparation called *pinole*. Because they swell in the stomach, a small quantity of seeds will satisfy hunger.

Once triggered, germination (like birth) is an irreversible process, and when a seed embarks upon this road there is no going back. The timing of germination is therefore vital to survival, and seeds have evolved a whole range of tricks to help them get it right. Long-lived seeds get all the press attention, but there are species that are equally deserving of our notice because of the astonishing brevity of their lives. Poplar and willow seeds are very short-lived and must find wet mud within hours of dispersal or perish. The large seeds produced by many tropical rainforest trees decay within weeks if they have not germinated. For such species correct germination timing is really about getting the timing of seed production right, because once the seeds are mature they have a limited shelf life.

A few species are viviparous and have reduced the duration of the seed phase of the life cycle to its absolute minimum, letting seeds germinate while they are still in the flower head. This sometimes happens in ordinary wild grasses if the weather is very warm and wet. It can also happen in cereals such as wheat, at great loss to the crop.

Vivipary occurs in some mangroves, which are saltwater-tolerant trees that live along coastlines throughout the tropics. These trees drop their precocious offspring like torpedoes into the ocean, where they float upright, ready to sink themselves into the mud and take root as soon as they have been deposited by the tide. In some mangrove species the pregerminated seedling hides inside the fruit pod, primed and ready to grow, but protected inside its mother's vessel until the right moment.

Seed dormancy comes into its own in seasonal climates that are either too dry or too cold for seeds to germinate during part of the year, so that it pays to wait out the unfavorable season. The range of seed dormancy behaviors is truly vast. It has been said of dormant seeds that, to paraphrase Malvolio in Shakespeare's *Twelfth Night*, "Some are born dormant, some achieve dormancy, and some have dormancy thrust upon 'em." But such elegant distinctions are too simple to do justice to the subtleties and sophistication of seeds. Indeed, it is almost as though every seed has a personality all its own, partly inherited, partly a characteristic of its species, and partly a product of the particular environment and history that it has experienced, but all ultimately shaped by, and subject to, natural selection.

A simpleton in the spectrum of sophistication is knotgrass *Polygonum aviculare*, a weed of arable land and gardens, whose seeds are born dormant and will not germinate before winter. Chilling of seeds in the soil during the winter months breaks their dormant state and prepares them to germinate when the soil warms up in spring. Any seeds that have not germinated by May gradually become dormant again and require another period of chilling before they are once again ready to germinate. To call knotweed a simpleton for exhibiting such rational behavior is not really fair, particularly since it is in good company. Many weed seeds undergo similar cycles of dormancy and germinability with the seasons, but it is certainly at the less sophisticated end of the spectrum of seed behavior.

One rung up the ladder is another very common weed, fat hen *Chenopodium album*. This plant produces nondormant seeds early in the year and dormant ones in the main crop. This enables fat hen to get an extra generation into the year with the seeds produced early, while the later-produced ones are held safely over to the following spring. Even more sophisticated are the annual plants that make a bet on the severity of winter. These so-called winter annuals, like thale cress *Arabidopsis thaliana*, cheat grass *Bromus tectorum*, and numerous others, germinate in autumn. This runs the risk that winter frost will kill the seedlings, but the payoff comes from those seedlings that survive because they have a head start in growth compared to seeds that wait till spring to germinate. If they survive, they are bigger and produce many more seeds.

Winter annuals usually hedge their bets by keeping some seeds dormant over winter. These spring-germinating seeds produce smaller plants and have fewer seeds than the survivors of the winter cohort, but their chances of survival are usually higher.

Because winter is severe in some years and less severe in others, sometimes it's the nondormant seeds that do better and sometimes it's the dormant ones. Simple calculations have shown that winter annuals that hedge their bets by producing a mixture of the two seed types do much better in the long run than plants that produce only one seed type or the other. It's simple economics that you should spread your bets against the risks created by uncertainty, but not even the doyen of long-term investors Warren Buffett can claim, as plants quite justifiably could, that he has successfully played this strategy for millions of years.

So far, we are still less than halfway up the ladder of sophistication for germination behavior. We have seen how plants can modulate dormancy in line with the seasons and we have witnessed bet hedging. All these plants use the season, temperature, and soil moisture to program seeds for dormancy and to trigger them into germination at the right time, but seeds use other cues to give them more precise information on when to germinate. Many seeds, like lettuce *Lactuca sativa* for example, are light-sensitive and will not germinate if kept in the dark, even when the temperature and moisture are right. This mechanism prevents seeds from germinating when they are too deeply buried in the soil to have any chance of reaching the surface. Even the briefest glimpse of daylight is sufficient to release such seeds from dormancy. Digging over a vegetable bed flashes a signal to huge numbers of light-sensitive weed seeds that they are near the surface. This is why turning over the soil in your garden should be avoided as much as possible if you want to minimize the amount of weeding you have to do.

Many seeds have an even better trick up their sleeves. Plants perceive light by means of an anciently evolved photoreceptor molecule called phytochrome. The phytochrome molecule exists in two forms that are mutually interconvertible. The form called P_r absorbs red light and in doing so is converted to the other form, called P_{fr}. This form of phytochrome is particularly sensitive to far red light (also called near infra red) which has a longer wavelength than red. Far red light causes P_{fr} molecules to convert to the P_r variety. What is the point of all this molecular to-ing and fro-ing? Because the two forms of phytochrome are sensitive to light of different wavelengths and are interconvertible, the ratio of P_r to P_{fr} is determined by the relative amounts of light that the plant is receiving at the two wavelengths. This ratio contains information about the local environment that is of vital importance to the plant.

Unobscured daylight has a red/far-red ratio of about 1, which results in a balanced ratio of P_r to P_{fr} in an illuminated plant. However, when sunlight passes through a leaf, the red light is preferentially absorbed. The remaining light passing through the leaf (or reflected from it) is depleted in red, which is why plants appear green. Light that has passed through a leaf therefore has a red/far-red ratio much less than 1, and this is detected by plants through its effect on the $P_r:P_{fr}$ ratio. Plants use phytochrome to sense where their neighbors are and to adjust their own growth so as to avoid them. Many seeds also use phytochrome in the same way and will germinate in darkness, but not if exposed to light that has passed through a leaf. Smaller plants invariably lose any contest with others that are large enough to shade them, so it is better for a seed to remain dormant than to germinate beneath the shade of another plant where the seedling will only struggle and die.

A drawback of the phytochrome system is that it can't work in darkness, so only seeds lying on or near the soil surface can use it to detect whether potential competitors are present or not. However, there is another cue that buried seeds may use to detect whether there is a gap in the vegetation above their heads. Vegetation such as grass acts as an insulating layer over the soil surface, which moderates the range of temperatures experienced by the soil beneath. Bare soil has no such insulating layer and so seeds buried beneath it experience extreme fluctuations of temperature. It turns out that many buried seeds use these fluctuations as a signal that the soil surface is bare. They will germinate in spring if they have experienced fluctuating temperatures, while a seed that has experienced the same average temperature, but without the fluctuations, will not germinate.

There is one group of species whose seeds benefit from the presence of other plants: parasites. The witchweeds *Striga* spp. are flowering plants that live by parasitizing other species. *Striga* seeds are stimulated into germination by chemical substances exuded from the nearby roots of their host plants. The seeds of the parasite are tiny, exceedingly abundant, and viable in the soil for up to twenty years, making control very difficult. *Striga* species are especially serious weeds in African crops belonging to the grass family (corn, sorghum, millet, rice, and sugar cane) and also in crops of the legumes cowpea, peanut, and soybean.

However, it has been discovered that two legumes, silverleaf *Desmodium uncinatum* and greenleaf *Desmodium intortum*, can actually be used to control at least one of the most serious *Striga* species by exploiting the germination mechanism of the parasite to deliver it a sucker punch. The roots of the *Desmodium* species exude two substances, one that stimulates *Striga* germination and another that inhibits germinated seedlings from forming the special kind of roots called haustoria through which the parasite beggars its host. These molecules are examples of the subtler weapons in the chemical battle between plants and their enemies. More potent poisons yet fill the well of sorrow's mysteries.

{ 13 }

Sorrow's Mysteries

POISONS

No, no! go not to Lethe, neither twist
Wolf's-bane, tight-rooted, for its poisonous wine;
Nor suffer thy pale forehead to be kiss'd
By Nightshade, ruby grape of Proserpine;
Make not your rosary of yew-berries,
Nor let the beetle, nor the death-moth be
Your mournful Psyche, nor the downy owl
A partner in your sorrow's mysteries;
For shade to shade will come too drowsily,
And drown the wakeful anguish of the soul.
JOHN KEATS, FROM "ODE ON MELANCHOLY"

On September 11, 1978, Georgi Markov, a Bulgarian exile living in London, was on his way to the offices of the BBC where he worked as a journalist when, as he approached the queue for the bus, he felt a sudden stinging pain in the back of his right thigh. Turning, he saw a man bending to retrieve a dropped umbrella. Speaking with a foreign accent, the man apologized, hailed a taxi, and was gone. Though in pain, Markov caught his bus and continued to work, but by evening he had developed a high fever and was taken to the hospital, where he was diagnosed with blood poisoning. Markov failed to respond to treatment and within three days he was dead. Georgi Markov was an outspoken critic of the Bulgarian Communist government, broadcasting to that country on the World Service of the BBC and other radio stations. He had survived two previous attempts on his life, one of them by poison, but it was evidently third time unlucky.

At the autopsy, pathologists found what they thought was the head of a pin in Markov's thigh, but as they tried to extract the pin it tumbled onto the tabletop, where they could see that it was in fact a tiny pellet. Under the microscope the pellet was found to have tiny holes drilled into it, making a chamber that could have contained poison. Though the chamber was empty, only one poison delivered in this way in such a tiny amount, less than one two-thousandth of a gram, could have killed Markov: ricin. Some years later, two defectors from the Russian KGB publicly admitted that the organization had used an agent to assassinate Markov with ricin. Ricin is found in castor bean seeds and is a powerful inhibitor of the machinery that makes proteins in every cell of the body. Ricin is more toxic than cobra venom and has no known antidote.

Cyanide is another seed poison found in a wide range of plants and occurs in the pits of fruit such as apples, cherries, and almonds. Indeed, cyanide has the odor of burnt almonds. The notorious poison strychnine is also found in seeds contained within an otherwise harmless fruit. The castor oil plant (*Ricinus communis*) is a tropical species, but it is commonly grown as a houseplant, and the polished, attractively patterned beans are sometimes strung as ornaments on necklaces and sold to tourists. Chewing a single bean can kill a small child. The seeds of laburnum (*Laburnum anagyroides*) are another commonly recognized hazard that causes accidental poisoning. In the same plant family as laburnum, red kidney beans are poisonous if not cooked at a sufficiently high temperature, which is why they should never be prepared in a slow-cooker. This device applies a low heat for a prolonged period, which will cook food but is a recipe for disaster where kidney beans are concerned.

Why are some seeds poisonous? The answer is somewhat obvious if you think about this question from the perspective of the plant. Seed poisons defend a plant's immature offspring against being eaten by animals. Ricin is just part of a castor bean's armory, the main part of which is to lock its food stores up in an indigestible type of oil. Wild oilseed rape (or canola, *Brassica napus*) has a similar trick, and 40 percent of the oil in its seeds is a fatty compound called erucic acid, which is harmful to the heart when fed experimentally to laboratory rats. Strains of *Brassica napus* bred for oil production have lower amounts of erucic acid that can be reduced further during processing.

If it is obvious why some seeds are poisonous, the really interesting question is why so many are not. One answer is that seeds can be protected in other ways, as nuts are by heavy armor, or peanuts by burial, but there is another reason too. The relationship between seeds and animals is not simply one of prey (the seed) and predator (the seed eater), but is often more complicated than that. Squirrels, for example, eat nuts, but they also cache a proportion of them away for later consumption in times when seeds are in short supply. Rodents and some seed-eating birds are therefore seed dispersal agents as well as predators and, from the point of view of the plant, the relationship has benefits as well as costs. Plants with palatable seeds sacrifice a proportion of them to animal predators in return for the remainder, though this may be only a very small fraction, being dispersed to places where they may germinate successfully.

Poisons like ricin tend to be highly specific in their biological action, interfering with life processes at very particular points. Ricin attacks a process that is so fundamental to life that all people are susceptible to it, but there are other cases where the specificity is so finely tuned that a compound may be poisonous to only a fraction of the human population who have a genetic susceptibility to it. Favism is an inherited condition mainly affecting individuals from populations in the Mediterranean and Africa, for whom the broad bean (*Vicia faba*) is toxic. Sufferers can eat other beans, peas, and lentils without harm, and broad beans are not toxic to nonsufferers. Favism is caused by an abnormality in the gene that produces a substance called G6PD, which, in its abnormal form, results in the destruction of red blood cells when broad beans are eaten. The symptoms include jaundice and can vary in severity, sometimes even being fatal. An interesting peculiarity of favism is that most sufferers are male. Favism, like the better-known but much rarer disease hemophilia, is a sex-linked condition. The reason for this is that the G6PD gene is carried on the X chromosome, of which males have only one copy. Females have two X chromosomes (one inherited from each parent), and only if both Xs happen to carry the abnormal gene will a female suffer from favism. A good copy of the gene on one X can compensate for a defective copy on the other.

The selectively poisonous effects of fava beans may explain why the ancient Greek philosopher and mathematician Pythagoras forbade his followers to eat beans. This was an especially curious prohibition because the Pythagoreans were vegetarians and beans are usually a very important source of protein in a meatless diet. Pythagoras and his followers lived in southern Italy, where favism occurs today, although it is a rare condition. It may well have been much more common two and a half thousand years ago in the days of Pythagoras, because the G6PD abnormality confers some protection against malaria. The disease would probably have been more common in Italy then than now. However, we shall never really know why Pythagoras wouldn't eat beans. Perhaps they just didn't agree with him.

The seed protein gluten, which gives dough made from strong wheat flour the wonderful elastic properties that are so valuable in making a good loaf or a light pizza base, is not normally considered a poison, but it can have devastating effects upon people with celiac disease if their condition is not diagnosed. This is another inherited condition that causes what is meat to one person to be poison for another. Ambrose Bierce summed up the dual nature of plant poisons in typically sardonic fashion when he wrote in his *Devil's Dictionary*: "Belladonna: In Italian, a beautiful lady; in English, a deadly poison." The reason that the plant in question (*Atropa belladonna*) is called belladonna is that an extract of the plant was formerly used as a beauty aid to dilate the pupils of women to make them appear more attractive.

From an ecological perspective, the specificity of seed poisons provides plants with the ability to pick and choose among types of seed predator, favoring some while deterring others. The alkaloid capsaicin, which makes chilies taste hot and which is concentrated in their seeds, is a deterrent to rodents, but it does not affect birds which disperse them. Chili seeds pass unharmed through the gut of a bird, but they would not survive the grinding action of rodent teeth if, like humans, they developed a perverse liking for them. Seeds of the neem tree from Burma contain selective poisons that are injurious to insects but harmless to birds and mammals. Extracts of neem seeds are successfully used as pesticides on crops.

The biological specificity of poisons in seeds and other plant parts means that many of them are of medicinal value when given in the right dose because they can be used to target particular malfunctions in the body. Ricin, for example, has been investigated as an anticancer agent because of its cell-killing properties. An herbal extract of grapefruit seed has antimicrobial properties, but is harmless to humans and animals.

Just as weapons change hands between people, poisons are traded between micro-organisms, plants, and animals, and not all of a seed's chemical armory need be homegrown. The strangest story of this kind has only recently unfolded on the Pacific island of Guam, where a fifty-year-old medical mystery was solved by ecological detective work. The indigenous Chamorro people of Guam have some unusual eating habits, one being an inordinate fondness for Spam (the processed meat, not unsolicited e-mail) and another a liking for tortillas made from fadang. Fadang is a flour made by grinding the seeds of cycads, which grow in great abundance on the island. Cycads are some of the earliest-evolved seed plants, with shiny, very stiff, prickly leaves that must once have proved a challenge for plant-eating dinosaurs. Around the middle of the twentieth century it was noticed that Chamorros were affected by a neurological disorder peculiar to Guam. The Chamorros called the disease *lytico* or *bodig*, depending upon the precise symptoms, but all forms progressively paralyzed and eventually killed the sufferers. In the 1950s 10 percent of adult deaths among Chamorros were due to lytico-bodig.

The neurologist and author Oliver Sacks visited Guam in the early 1990s to investigate lytico-bodig, which bore a resemblance to a condition he had studied in New York and described in his book *Awakenings*. Incidentally, Sacks also wanted to see cycads, which had been his lifelong passion as an amateur botanist. In his book *The Island of the Colour Blind*, Sacks describes his visit to Guam, the history of lytico-bodig there and the possible connection between the disease and poisons found in cycad seeds. The poison theory had waxed and waned in popularity among medical investigators through the 1960s, as first one and then another poisonous component of the chemical armory found in cycad seeds was identified and then eliminated from enquiries. First in the frame was a poison called cycasin that had a range of toxic effects, being a potent cause of cancer at low doses and causing acute liver failure at high doses. But cycasin does not cause nerve damage of the kind associated with lytico-bodig. Then a substance with the shorthand name BMAA was discovered. BMAA is very similar in structure to a neurotoxic substance found in seeds of the chickling pea (*Lathyrus sativa*). The chickling pea is poisonous when eaten in large amounts, often causing death in India when drought and poverty force people to resort to this plant as the only food available to them.

The neurotoxin in the chickling pea and BMAA belong to a group of natural chemicals called nonprotein amino acids. Nonprotein amino acids are poisonous because they are impostors for the essential building blocks from which proteins are built. Twenty-one varieties of amino acid are kosher constituents of proteins and determine how they function. A protein is rather like a Lego model built from a range of twenty-one types of brick. Amino acids can be put together in uncountable numbers of ways to do everything from making the enzymes that digest the contents of your stomach to building the muscle proteins that power your heart or the nerve cells that make it possible for you to read the words on this page. Nonprotein amino acids are like Mega Bloks in your Lego box. The different brands do not fit together properly. Big Lego structures can be fatally weakened by an admixture of just a few Mega Bloks. So finding BMAA in cycad seeds that Chamorros use to prepare tortillas seemed like finding a smoking gun for the cause of lytico-bodig. But there were some pieces of evidence that didn't quite fit. Chamorros carefully wash fadang three times during its preparation to remove poisons. Also, none of the other neurotoxins known at that time, like the nonprotein amino acid in the chickling pea, had delayed effects. Lytico-bodig could develop decades after any possible exposure to fadang, and this was quite unlike other diseases caused by neurotoxins.

By the 1990s when Sacks visited Guam, lytico-bodig was disappearing, although why this was happening was as great a mystery as what caused it in the first place. Cycad poisoning and half a dozen other possible causes of the disease had been investigated and found wanting, and Sacks left Guam resigned to the prospect that lytico-bodig might disappear altogether before its cause would be tracked down. A fortunate thing for the

Chamorros, but hardly a triumph for medical science. That is where the story, as told in *The Island of the Colour Blind*, ends, but there is a sequel with an ending as unexpected and satisfying as the plot of any whodunit.

Elsewhere in the Pacific, the American botanist Paul Alan Cox, working at the National Tropical Botanic Garden in Hawaii, was also thinking about the connection between cycad poisons and lytico-bodig. He thought that BMAA might have been finding its way from cycad seeds into the Chamorro diet through another route, not involving fadang. He told me how the idea came to him. "I was sitting on the beach in Samoa when I realized that a toxin could potentially be biomagnified by flying foxes feeding on cycad seeds." Flying foxes are bats, and biomagnification is a process by which certain substances become concentrated as they move up a food chain. Cox investigated his hypothesis and tried to disprove it: " For example, we knew from 30 years of National Institutes of Health epidemiology that lytico-bodig occurred at a higher rate in men than women." So they asked Chamorro men and women about their bat consumption. "We found that flying fox eating is almost like a rite of passage in Chamorro men, but that women don't like to eat the flying foxes as much—they think they look like rats."

We had other ways to try to disprove the hypothesis, but couldn't. I still lacked confidence in putting it forward until one day I was showing a group around our Kauai gardens, when Marcia Williams, who was in the group, got interested in a cycad tree. When she heard me tell the story, she asked me if I had contacted Oliver Sacks. I explained to her that obscure botanists like me who are exiled to remote rocks in the ocean do not call world-famous scientists like Oliver Sacks out of the blue. She then told me that her husband, Robin Williams, had portrayed him in the movie version of Sacks's book Awakenings. *The next week I received a note from Dr. Sacks inviting me to see him next time I was in New York. I went with great trepidation, but when he expressed excitement about the hypothesis, I asked him to publish with me.*

The result was a joint publication by Cox and Sacks outlining what Sacks liked to call Cox's "batty hypothesis."

To test the batty hypothesis, Cox and colleagues analyzed museum specimens of flying foxes from Guam for BMAA and found huge amounts in the tissues. Flying fox tissue contained concentrations of BMAA that were a thousand times greater than in fadang flour. You would have to eat more than a ton of fadang to ingest the amount of BMAA found in a single flying fox. In fact, flying foxes contained a hundred times more BMAA per gram of tissue than is found in fresh cycad seeds. But where did it all come from? It was discovered that the ultimate source of BMAA is a microorganism known as a cyanobacterium that lives inside specialized roots in cycads. Unlike normal roots, those containing cyanobacteria are found on the surface of the soil where the inhabitants of the root have access to light. Cyanobacteria benefit the cycad by performing a trick that no plant or animal is able to perform for itself. They turn atmospheric nitrogen gas, which plants cannot use, into soluble nitrogen compounds that feed the host plant. It appears that cycads receive not only nitrogen from their microbial guests, but also a useful chemical weapon in the form of BMAA. The amount of BMAA in a gram of cyanobacteria is relatively small, but this is concentrated a hundredfold in the cycad and a hundredfold again in flying foxes. Thus, BMAA is biologically magnified ten thousandfold as it passes through the Guam ecosystem. At the top of the food chain are the Chamorro people who eat flying foxes whole, cooked in coconut milk.

All the pieces of the lytico-bodig mystery now fell into place. Further research by Paul Cox and his team shed light on two features of lytico-bodig that have never been properly explained before. The recent disappearance of the disease from Guam coincided with the extinction of one species of flying fox and the near extinction of another due to hunting. The disease is disappearing because the lethal delicacy that causes it is no longer available to the Chamorro people. And the delayed onset of symptoms may be explained by the discovery that BMAA can accumulate in a hidden reservoir, bound to proteins from which it may be slowly released in the brain. High concentrations of protein-bound BMAA were also found in fadang, so this flour may after all be a source of the poison, although if it were a major source, presumably lytico-bodig would not be disappearing with the flying fox. Finally, Cox's team looked for BMAA in its free and bound forms in human brain tissue. They compared samples from patients who had died of lytico-bodig with samples from two Chamorros who had died from other causes. BMAA was found in all the lytico-bodig patients and also in one of the two other Chamorros. Samples from two groups of Canadians were also analyzed: one group of two Canadians had died of Alzheimer's disease, which shares some symptoms with lytico-bodig, and the other group of eleven had died of non-neurological causes. There was no BMAA in the group of eleven, but it did turn up in the brains of the two victims of Alzheimer's. These patients might have acquired BMAA from some other environmental source, as cyanobacteria occur in many aquatic and terrestrial ecosystems. How many cases of Alzheimer's disease around the world are due to neurotoxins acquired in the diet is simply not known, but it is a possibility that might never even have been uncovered, if not for Paul Cox's batty hypothesis.

{ 14 }

Ah, Sun-flower!

OIL

Ah, Sun-flower! Weary of time,
Who countest the steps of the Sun,
Seeking after that sweet golden clime
Where the traveller's journey is done.
WILLIAM BLAKE, FROM "AH! SUN-FLOWER," *Songs of Experience*

The restless bud of the sunflower turns with the passage of the sun across the sky, warming its developing seeds before its petals unfurl and the flower head comes to rest facing east, its flaming yellow rays a refulgent image of the rising sun itself. The poetically cherished notion that the sunflower's open flower head continues to track the sun back and forth across the sky is widespread, but sadly wrong. Only the unopened bud tracks the sun, though countless poets have exercised their license to imagine otherwise. Scientific observation may have robbed sunflowers of poetic motion, but it has compensated with a flattering amount of attention to their evolutionary origins.

Sunflowers are indigenous to North America and their seeds

were collected and cultivated by Native Americans. In 1875 John Wesley Powell described seed harvesting and preparation by a tribe in the Grand Canyon region:

They gather the seeds of many plants, as sunflowers, golden rods and grasses. For this purpose they have large conical baskets, which hold two or more bushels. The women carry them by their backs, suspended from their foreheads by broad straps, and with a smaller one in the left hand, and a willow woven fan in the right, they walk among the grasses, and sweep the seed into a smaller basket, which is emptied, now and then, into a larger, until it is full of seeds and chaff; then they winnow out the chaff and roast the seeds. They roast these curiously; they put the seeds, with a quantity of red hot coals, into a willow tray, and, by rapidly and dexterously shaking and tossing them, keep the coals aglow, and the seeds and tray from burning. As if by magic, so skilled are the crones in this work, they roll the seeds to one side of the tray, as they are roasted, and the coals to the other. Then they grind the seeds into a fine flour, and make it into cakes and mush.

Native American tribes cultivated as well as gathered sunflower seeds. Domesticated plants with large heads, like those of modern cultivars, were widely distributed across North America before Europeans arrived. These cultivated sunflowers have all the other hallmarks of domestication that have been found in crops such as barley, wheat, and corn. Their stems do not branch, their seeds are much larger than those of wild relatives, seeds do not fall from the seed head when ripe, and they lack seed dormancy.

When he wrote his classic book about the plant in 1976, the dean of sunflowers Charles B. Heiser Jr. had little doubt that sunflowers had the distinction of being the only New World crop to have been domesticated North of Mexico. But how it had happened was a bit of a puzzle. The wild plant is found in the Southwest, while the earliest archaeological deposits containing cultivated sunflower seeds (too large to have been wild in origin) were all in the central and eastern areas of the continent. Cultivated sunflower seeds 3,500 years old were found in an excavation in Mammoth Cave National Park in Kentucky, for example. Weedy sunflowers, intermediate in size between wild and cultivated forms, also occur in the central and eastern United States today. Heiser's explanation was ingenious:

As man began to use the plant [in the Southwest], seeds were carried from place to place. Seeds scattered accidentally might find the new habitats around Indian villages suitable for their growth. Thus the plant may have become a camp-following weed, and was spread to new areas as the Indians moved. Although the plant was not yet a specific object of cultivation, the first close link to man had been made. We do not know the exact area or how large an area the common wild sunflower occupied before man came, but it certainly seems that the plant was spread by man into many new areas from its original home in the Southwest—west to California, south to Texas, and east across the Mississippi River. . . . gradually there began to evolve a special weed in the eastern area of its range. The most significant feature about this new plant was that it could grow only in the disturbed sites around Indian villages, and it also had larger heads and consequently larger achenes, which meant that it was an even better food plant than the original wild type. Today this eastern weedy form of the common sunflower is recognized as distinct from that of the western United States, although the two intergrade over a broad area.

Achenes are the botanical name for sunflower "seeds," which, technically, are dry fruit.

So, Heiser conjectured, the sunflower followed Native American tribes eastward, evolving a dependence upon humans along the way until it was fully domesticated in areas where wild sunflowers were not available to compete with it. That is where the story rested, until in 2001 eastern North America's claim to have the earliest domesticated sunflower was trumped by a find in Tabasco, on the coast of Mexico. A large sunflower seed, of a size typical of cultivated varieties, was discovered in archaeological excavations and dated to more than 4,000 years before the present. It looked as though Mexico, the undisputed home of corn, beans, chilies, and squashes, had just robbed the United States of its main claim to be a hearth of domestication—an irony that no doubt the migrant workers from Mexico who cross the border in their tens of thousands to harvest crops in the United States today would relish. This, however, was not to be the end of the story.

The study of crop domestication is not the province of a single branch of science, but the progeny of many. Archaeology finds the plant, botany identifies it, physics dates it with radiocarbon and other methods, and genetics sketches the evolutionary tree. Sometimes it is possible to retrieve DNA for genetic studies from archaeological specimens, but only if they have been preserved in unusually good condition. Because seeds of cultivated plants are such rich morsels for animals and microbes alike, they are most often preserved only as inedible charred remains that contain no surviving DNA. However, there is another way to use genetics to trace the origins of plants, and that is to test which living varieties are closest relatives of one another.

In 2004 a study of this kind compared the DNA of sunflower cultivars with that of wild populations of *Helianthus annuus* collected in the United States and Mexico. All the cultivars tested (including the varieties USDA and Mammoth; other plants traditionally grown by the Hopi, Havasupai, and Seneca tribes in the United States; and cultivars grown in Mexico) were found to be more closely related to North American wild populations than to Mexican ones. This proves that the sunflowers cultivated today descend from North American populations rather than originating in southern Mexico. Not only that, but among the twenty-one wild populations tested, the one most closely related to the cultivars was from Tennessee, in the eastern United States. Exactly as Heiser speculated, it appears that the sunflower was domesticated from weedy stock in the East.

But what of the 4,000-year-old Tabasco sunflower? That must either have been the product of a separate domestication event that went extinct, or possibly even a descendant of the eastern North American domestication that was carried south. Seeds, after all, are designed by nature to travel, and as the later history of the sunflower shows, the plant was only just beginning its journey around the world.

Although European settlers in the United States rapidly adopted corn, another Native American crop, they did not grow sunflowers on any scale until the second half of the twentieth century. Strangely, like a prophet without honor in his own country, it was not in the United States that the potential of sunflower seeds as a crop was first recognized. Well might the sunflower head, pregnant with seeds, fix its gaze eastward, because it was the Russians who adopted the plant, bred and improved it as a crop, and were the source of its spread to other parts of the world, including its eventual return to widespread cultivation in North America. The varieties Mammoth and USDA are both from Russian stock.

Perhaps the sunflower was ignored in the United States because it was too familiar, but in Russia it was adopted for precisely the opposite reason—it was almost unknown. In the early nineteenth century the Russian Orthodox Church issued a holy decree that proscribed a list of oil-rich foods from being eaten during Lent or in the forty days before Christmas. These two periods fall in the coldest months of the year when rich food is particularly comforting and sought-after, but almost anything with a high oil content was forbidden. Sunflower seeds, containing about 30 percent oil, were so little known in Russia at the time that they were not named on the list. Forthwith, sunflower seeds and their oil were eagerly adopted in Russia, without fear of religious disapproval.

As the Russians eager for a source of fat discovered, it can pay to think unconventionally. So imagine a chemical factory capable of turning water, manure, and air into high-quality, high-value oils and which runs entirely on solar power, emitting only oxygen as a waste product. The oils produced by this factory are hygienically packaged in a dry form that is easily transported, stored, and processed, and the empty packaging can be recycled as a nutritious food that can be fed to livestock. Manure from the livestock can be sent back to the factory to help make more oil. If it existed, how much would such an environmentally friendly factory system be worth to an industrialist? Billions, surely. Of course such factories do exist and they are dirt cheap or even free to any entrepreneur who wants to pop a sunflower seed into the ground and watch it produce a hundred more by the end of the season.

Sunflower oil is just one of hundreds of product lines that oil-producing plants of the botanical kind are capable of producing. The worldwide trade in seed oils is worth more than $61 billion a year. The range of oils from seeds and their variety of uses are truly astonishing. Candlenut (*Aleurites molucanna*), which the early Polynesian settlers brought with them to the Hawaiian Islands, is so oily it will burn like a candle for three quarters of an hour. To tell the time, several nuts would be impaled on the midrib of a palm leaf and the nut at the top lit. No doubt, "Be home by the third nut," was a commonly heard domestic command in Polynesian households.

Seeds of the castor bean *Ricinus communis*, the source of the notorious poison ricin, may contain over half their weight in castor oil. This oil has been used to fuel lamps since ancient times and has the advantage that it does not become rancid in storage. Cold-pressed castor oil does not contain ricin, but if consumed it causes nausea, vomiting, and diarrhea. It is a drastic but effective remedy for constipation. Castor oil and derivatives from it have a hundred and one other commercial uses in cosmetics, pharmaceuticals, adhesives, and explosives, as a plasticizer, and as a lubricant in jet engines. Not surprisingly, the U.S. Congress has included castor oil as a strategic resource of importance to the defense of the nation.

The seeds of certain palms are some of the richest sources of oil. That from the seed of the African oil palm (*Elaeis guineensis*) is edible and is also used in the manufacture of soap, hence the brand name *Palmolive*. Chocolate is made with a fat known as cocoa butter that is extracted from beans of the cocoa tree (*Theobroma cacao*). If it was recently printed, you may detect the pleasant odor of linseed oil from the ink on the pages of a book, perhaps even this one! Oil paints are also made with linseed oil, as is putty used in glazing. The oil of the desert jojoba bush is used in cosmetics. Soya, sunflower, and oilseed rape (or canola), peanut and corn (maize) are all fast-growing crops which yield inexpensive oils that are used in foods and cooking.

In 2002, supermarkets in Wales were surprised by streams of customers buying trolley-loads of cheap cooking oil. Because it was much cheaper than diesel, people were running their cars on canola (rapeseed) oil. This use was illegal because it evaded the tax on fuel, though in Germany it was allowed and even encouraged. Biodiesel, a mixture of refined petroleum and oil from seeds, is widely sold as a "greener," more climate-friendly, fuel. European Union regulations now encourage the addition of vegetable oils to fuel. While this may in principle sound environmentally beneficial, it has had an unintended environmental impact. Rainforest in Southeast Asia is being cleared for oil palm plantations that are destined to feed the new European market for "sustainable" fuel. The destruction of rainforest not only threatens biodiversity, but also releases CO_2 through tree burning and the oxidation of organic matter in soils. Biodiesel from oil palm plantations is therefore neither sustainable nor green. Exxon (Esso in Europe) used to advertise its gas with the slogan "Put a tiger in your tank." If you run your car on fuel from Southeast Asian palm oil plantations, you might as well display a bumper sticker that says "Put an orangutan in your tank," because rainforest destruction spells extinction for these apes.

One reason that oil palm is considered a good source of fuel is that it produces an especially energy-rich oil. Plant and animal oils and fats have the same basic chemical structure, but they vary in the amount of energy they store. Chemically, they are all triacylglycerols. The amount of energy stored in a triacylglycerol molecule depends upon its chemical structure. All triacylglycerols have a backbone of three (hence *tri-*) chains of carbon atoms, tethered to each other at one end like the prongs of a trident that lacks a handle. Each prong of the trident is a fatty acid, and the tether holding the three prongs together is a molecule of glycerol. The properties of any particular triacylglycerol depend mainly upon the number of carbon atoms in each fatty acid chain and how the carbon atoms in the chain are bound to each other. A carbon atom may be linked to its neighbor by one chemical bond (a single bond) or by two (a double bond).

If you are even the slightest bit diet-conscious (and who isn't?), then you will probably be familiar with the terms "saturated fat" and "polyunsaturates." In food advertisements the terms normally appear as "low in saturated fat" or "high in polyunsaturates," a clue to which is good for you and which is not. These terms refer to the chemical structure of the fatty acid chains. In saturated fats, the fatty acid chains all have single bonds in their carbon chains, while in unsaturated fats there is one or more double bond. Polyunsaturates have several double bonds.

Longer fatty acid chains and fewer double bonds make the triacylglycerol molecules stick to each other more, which gives the oil a higher melting point. Chocolate, which is rich in saturated fats, is a familiar example that is a solid at twenty degrees Celsius. Shorter chains and more double bonds in the carbon chain result in triacylglycerols with a lower melting point. Cooking oils, normally made from seeds containing unsaturated triacylglycerols, are examples that are liquid even if kept at five degrees Celsius in the fridge. Triacylglycerols that are solid at room temperature are commonly called fats, those that are liquid are referred to as oils. Sunflower oil is turned into solid margarine by chemically converting the double bonds of the unsaturated triacylglycerols in the oil into single bonds.

The oils in seeds are invariably mixtures of many different triacylglycerols, some saturated and others not. However, it is odd that seeds should contain any unsaturated triacylglycerols at all. There are two reasons why this is surprising. Firstly, saturated triacylglycerols, like those in palm oil, are more energy-rich than unsaturated ones. So why shouldn't every mother plant be as generous to her offspring as the oil palm is to hers and supply every seed with the most energy-rich food she can? You might think the answer is obvious—not every mother can afford the cost, but actually that cannot be the reason. All the triacylglycerols a plant produces are first synthesized in saturated form and only afterward are some of these converted to unsaturated molecules. This makes an unsaturated triacylglycerol molecule *more expensive* for a plant to manufacture than its saturated equivalent. So even though saturated triacylglycerols are more energy-rich than unsaturated ones, they are cheaper to produce in energy terms. That is the second reason it is odd that plants put unsaturated triacylglycerols in their seeds: they are expensive.

If plants store unsaturated triacylglycerols in their seeds despite the increased cost and decreased energy content of this oil, then there must be some advantage to doing so. What can it be? Of course it cannot be the same reason that humans favor unsaturated fats over saturated ones. Plants don't get heart attacks. Interestingly, there is a very clear geographical pattern to which plants have seeds high in saturates, and which ones have more unsaturated triacylglycerols. This gives us a clue to the advantage of unsaturated oil. The pattern is: the cooler the climate during the period of seed germination, the greater the proportion of unsaturated triacylglycerols in the seed. This pattern is so general, that it can even be found among different sunflower species. Wild sunflower populations native to Canada have only about 6 percent of their seed oil in saturated form. Twenty degrees of latitude further south, in Mexico, native sunflowers have 12 percent, or twice the proportion of saturated oil. This contrast can also be seen within wild populations of a single species like *Helianthus annuus*. A study found that in Texas 12 percent of the oil in seeds of this species was saturated, while in Saskatchewan, Canada, the proportion was less than half that.

What is the explanation for this pattern? The answer seems to be that at lower temperatures, seeds whose oil stores are held in saturated form have difficulty germinating. This is probably because most biochemical reactions take place in solution, so at low temperatures saturated oils are not liquid enough for germinating seeds to use them. Unsaturated oils, on the other hand, are apparently easier to metabolize at lower temperatures, as you might expect from their lower melting points. Although the oil in seeds appears to be in solid, not liquid form, oils and fats have the unusual property of behaving like liquids at the molecular scale when the temperature is as low as fifty degrees Celsius below their actual melting point.

So mother knows best after all! There would be little point in her providing her offspring with the equivalent of solid, energy-packed Mars Bars if the cool climate makes these snacks literally too hard to consume. Cool germination conditions call for something unsaturated. Plants that live in some of the coolest temperate climates, such as oats, barley, wheat, oaks, chestnuts, and beech, don't even use unsaturated oil, but store most of their energy as carbohydrate (starch). Oil is also only used by plants for energy storage when, as is the case in seeds, it needs to be packaged in concentrated form. Where bulk is not an issue, plants store energy in their vegetative parts as carbohydrates, such as the starch found in potatoes. Even in tropical plants like taro, the energy store in its corms is starch. So oil production is a speciality of seeds.

{ 15 }

John Barleycorn

BEER

There was three Kings into the east,
Three Kings both great and high,
And they hae sworn a solemn oath
John Barleycorn should die.
They took a plough and plough'd him down,
Put clods upon his head,
And they hae sworn a solemn oath
John Barleycorn was dead.
But the cheerfu' Spring came kindly on,
And show'rs began to fall;
John Barleycorn got up again,
And sore surpris'd them all.
ROBERT BURNS, "JOHN BARLEYCORN"

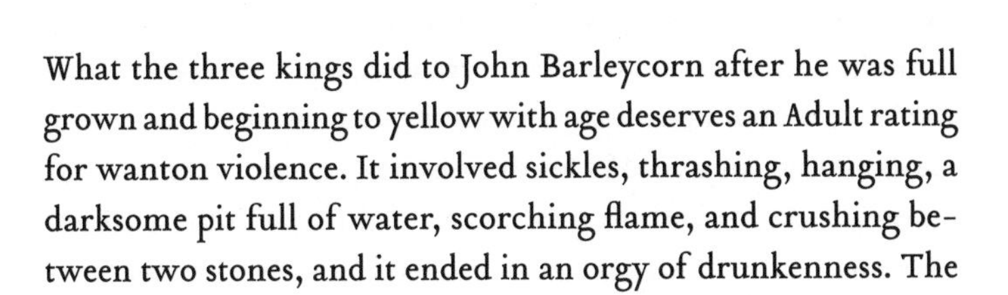

What the three kings did to John Barleycorn after he was full grown and beginning to yellow with age deserves an Adult rating for wanton violence. It involved sickles, thrashing, hanging, a darksome pit full of water, scorching flame, and crushing between two stones, and it ended in an orgy of drunkenness. The bowdlerized version of the recipe for making beer from barley is less entertaining but more enlightening. Barley seeds are first sprouted so that the enzymes which become active upon germination convert the starch stored in them to the sugar maltose; roasting the sprouting seeds converts some of the sugar to malt, and then the mash is fermented to convert the sugar to alcohol.

The importance of beer to human health and happiness should not be underestimated. In the days before clean water supplies, "small beer" (beer with low alcohol content) was a safer drink than polluted well water. The Roman historian Pliny the Elder commented in his monumental, twenty-four-volume work *Natural History* that "the nations of the West also have their own intoxicant, made from grain soaked in water. There are a number of ways of making it in the various provinces of Gaul and Spain. . . . Alas!, what wonderful ingenuity vice posses! A method has actually been discovered for making even water intoxicated."

Pliny's remarks betray the disdain for beer of a patrician wine drinker commenting on the uncivilized customs of barbarians who lived at the far-flung reaches of the Roman Empire. Pliny was less disparaging of barley as food and wrote that it was ancient and venerated by the Athenians. Even in Rome, gladiators were once known as *hordearii*, or "barley eaters."

Archaeology proves Pliny right about the ancient role of barley in the human diet. Barley was important in the transition from hunter-gathering to settled agriculture and was one of the first three grain crops to be domesticated at the dawn of Old World agriculture in the Fertile Crescent. This region, curved like a sickle blade, sweeps in a broad arc from Israel and Palestine on the eastern shores of the Mediterranean, northward into Syria and Turkey and then southeast into the valleys of the Tigris and the Euphrates in Iraq. The seeds of barley, preserved as charred remains, tell the story of how the Neolithic revolution, which gave birth to settled agriculture, probably started. Excavations at the Neolithic settlement of Netiv Hagdud, near the town of Jericho, have uncovered one the earliest episodes in this story. Netiv Hagdud was occupied for only three hundred years, ending about 8500 BC, so its early remains were not obliterated by later occupation. The people who lived there hunted a great variety of wild animals and also harvested grains using flint sickle blades that have been found at the site. The variety of wild plant foods collected was large, including figs, pistachios, acorns, and almonds, but the abundance of remains at the site suggest that barley was the staple food plant.

Were the barley grains found at Netiv Hagdud from domesticated crops, or were they collected wild? The answer can be found by microscopic examination of the plant remains themselves. The seeds of grasses stay attached within the flower as they develop, forming a unit called a "spikelet." The seed heads, or "spikes," of wild grasses such as wild barley (*Hordeum vulgare* subsp. *spontaneum*) shatter and disperse their spikelets as they ripen, leaving a scar where the spikelet detached cleanly from the head. By contrast, the spikes of domesticated cereals retain their spikelets and do not shatter, but are removed after harvest by threshing, which leaves diagnostic signs of breakage from the spike. When examined, the vast majority of barley remains found at Netiv Hagdud showed signs of shattering, rather than threshing, indicating that the grains were harvested from wild populations.

Two genes control whether barley shatters or not, and a small percentage of plants in wild populations are nonshattering genetic variants. Plants like these were crucial to the domestication of barley, which the archaeological record suggests may have begun around the time that Netiv Hagdud was abandoned. The transition from shattering wild barley to a nonshattering domesticated crop took no more than three centuries. What must have happened is a typical example of the process known as artificial selection, in which evolutionary change in an animal or plant population is driven by human preference, intentional or not, for particular characteristics. The different breeds of dog, all descended from a common ancestor, are a familiar example.

In the case of the domestication of barley, the preferences of early farmers were possibly unwitting because harvesting wild cereals with sickles would anyway have preferentially collected the seeds of nonshattering plants. Imagine gathering the stems of wild grasses that are on the point of ripeness. The blow of the sickle would dislodge seeds from the top of the spike, where they ripen first, and these would be lost. The harvest you brought home would consist of a disproportionate amount of unripe seeds and ripe seeds from nonshattering plants. Now, use some of those seeds to replant next year and repeat. Each year the proportion of nonshattering plants will increase, to the point at which the crop is fully domesticated and cannot disperse its seeds unaided.

But this was just the first stage of domestication. Through artificial selection, domestication wrought other important changes upon wild barley in the Fertile Crescent. Barley spikelets occur in triplets, arranged alternately up the spike, but in the wild species only the middle grain in each triplet develops. In the mature spike this results in two opposite rows of spikelets on the spike, and the first domesticated barley also had this two-row structure. A single gene controls the fertility of the lateral spikelets in each triplet. Where there are genes, there are mutations, and at some point soon after two-rowed barley was domesticated, barley with a genetic mutation causing all three spikelets to be fertile came under artificial selection. This mutant has six rows of spikelets on the spike, making it three times more likely that it will be resown than two-rowed barley, without any need for conscious selection on the part of the sower. Why the same numerical advantage does not cause the six-rowed mutant to replace the two-rowed one in wild populations of barley is a puzzle. The mutant must possess some disadvantage under natural conditions which prevents natural selection favoring six-rowed barley.

The Neolithic domestication of barley involved at least one probable instance of conscious selection on the part of early farmers. The seeds of wild barley and most of its cultivated forms remain wrapped in a part of the flower after threshing. This hulled grain is used in brewing and for animal feed, but "naked" grain without the hull was preferred by traditional farming communities for use in food. The naked/hulled characteristic of barley seed is controlled by a single gene, so some small proportion of plants in wild and early domesticated barley crops must have produced naked grain, but naked, six-rowed barley soon became a distinct crop variety through artificial selection. Since the presence/absence of the hull around the seed does not become apparent till after threshing, it does not seem likely that the harvest process itself would favor naked seeds. Farmers must have chosen naked grains to resow.

Two other cereals, emmer and einkorn wheat, were domesticated in the Fertile Crescent at the same time as barley and showed parallel evolutionary changes from shattering to nonshattering spikes as a consequence of artificial selection. Lentils and peas are two other important seed crops that were early domesticates in the Fertile Crescent. In these plants too, artificial selection blocked the normal seed dispersal mechanism of their wild progenitors, producing plants whose seed pods remained closed at harvest time. The large grains and fat peas and lentils we enjoy today are the result of artificial selection for larger seed size, another evolutionary legacy of a trend started in the Neolithic.

There is not just one instance of domestication in the production of beer, but two. In addition to barley, another organism essential to beer production is, of course, brewer's yeast (*Saccharomyces cerevisiae*). This species is also used in wine and bread production, but how it was domesticated is only a recent discovery. Brewer's yeast, just like domesticated barley and grapes, must have wild relatives, but *S. cerevisiae* is not that common in the wild. The closest relatives of brewer's yeast found in nature occur in fermenting fruit, in the sugary sap of trees, and in clinical samples from patients with a compromised immune system.

Because it is rare in nature, some have suggested that "wild" populations of *S. cerevisiae* may be escapees from domestication. At first this may sound far-fetched, but feral cats and pigeons infest cities, so why not feral yeast? Imagine the implications: our immune system generally protects us from yeast infections, but patients with compromised immune systems are susceptible to infections of all kinds. When you take a friend in hospital a bunch of grapes, the general idea is that the patient eats the fruit, not that yeast in the bloom on the grapes eats the patient!

To investigate the origins of wild yeast, Justin Fay and Joseph Benavides of Washington University School of Medicine in St. Louis, Missouri, constructed an evolutionary tree for eighty-one yeast strains collected from domesticated and wild sources all over the world. An evolutionary tree is like a family tree, showing the relationships among people through their common ancestors. It answers the question "How far back do you have to go before a particular pair of individuals share a common ancestor?" For close relatives like brothers and sisters, the answer is just one generation. For first cousins it is two generations, for second cousins it is three generations and so on. One of the differences between an evolutionary tree and a family tree is that evolutionary trees stretch back over thousands, or many millions, of years and show how change occurred over time.

The evolutionary tree for *Saccharomyces cerevisiae* revealed several interesting things. First, it settled the question as to whether wild strains were descended from domesticated ones. It turned out that they were not. For example, of eleven yeast strains isolated from infections in immuno-compromised patients, ten were like wild yeasts and only one appeared to be related to a yeast strain from a vineyard. This news will no doubt come as a relief to anyone in the habit of taking grapes to the bedside of hospital patients. Yeasts isolated from wild sources such as tree sap were also found not to be descended from domesticated strains. In fact, quite the reverse.

The yeast strains with the deepest, and hence oldest, origin in the evolutionary tree were from African palm wine that is made from the sap of the oil palm. Because wild yeast occurs naturally on tree sap, this may be how it first became associated with humans—through the fermentation of palm wine in Africa. Of course, the human species also originated in Africa, which raises the tantalizing possibility that *Saccharomyces cerevisiae* was used for wine fermentation in Africa long before barley and other cereals became domesticated and beer or bread entered our diet. Cereals were domesticated in the Fertile Crescent around ten thousand years ago, well after *Homo sapiens* had spread out of Africa.

Fay and Benavides' evolutionary tree also showed that the yeast used today to produce rice wine (sake) in Japan and the yeast used to produce grape wine belong to separate branches of the tree and represent two separate domestications, one in the East and one in the West. This is not unexpected, given that rice was domesticated in the Far East independently of cereal domestication in the Fertile Crescent. In theory, it should be possible to date the earliest common ancestor of the Eastern and Western branches of the *S. cerevisiae* tree. But, in practice, this requires some large assumptions to be made about, among other things, the length of the average generation time in yeast. Generation time is measured by the number of years between an individual producing its first offspring and those offspring themselves starting to reproduce. In other words, it is the time from when you become a parent to when you become a grandparent. This varies among human populations, but averages twenty to twenty-five years. A yeast generation is about three hours.

Assuming a generation time of three hours, Fay and Benavides estimated the split between the Eastern and Western yeast to have occurred more than twelve thousand years ago, but they admitted that if they had underestimated the generation time of yeast, the split could have occurred more than a hundred thousand years ago. The earliest evidence of fermented drinks so far uncovered is from China, where residues from the insides of Neolithic pottery jars indicate that they once contained an alcoholic beverage probably made by fermenting rice, honey, and fruit. The jars were nine thousand years old, so the estimate of yeast domestication twelve thousand years ago is in the right ballpark.

However the dates for yeast domestication pan out, and estimates will certainly become more accurate with further evolutionary and archaeological research, it is clear that humans have brewed alcoholic beverages since at least the Neolithic, and probably for a lot longer than that. Why, though, does *Saccharomyces cerevisiae* oblige us by producing alcohol? Does this substance, which is toxic to most microbes, serve some purpose for yeast? Possibly, but there is another, simpler explanation too: maybe it just can't help but produce it.

As you will know if you have brewed your own beer or fermented your own wine, yeast will turn sugar into alcohol (ethanol) only if it is deprived of oxygen. Let air into the fermentation vessel and all the yeast will do is turn sugar into water, carbon dioxide, and more yeast cells. However, in anaerobic conditions yeast is unable to oxidize sugar fully, and the reaction stops at an incomplete stage with the production of ethanol. There is so much chemical energy still locked up in ethanol that it can be used to run an internal combustion engine.

When deprived of oxygen, and hence an efficient energy source, yeast cells divide much more slowly, so there is little doubt that the anaerobic conditions in which ethanol is produced are suboptimal. But that is not the end of the story, because evolution has a habit of turning adversity to advantage. The advantage to ethanol production is that it poisons other microbes, hence its well-known preservative properties. When vice admiral Lord Horatio Nelson lay dying from a sniper's bullet on the day of his famous victory at the battle of Trafalgar, he asked that his body not be buried at sea, as was naval custom. Legend has it that to preserve Nelson's body for the long passage home it was stored in a cask of Royal Navy rum. Unfortunately, when Nelson's ship reached Portsmouth the rum had vanished. Sailors had tapped the cask with a small hole and drunk the contents.

Yeast has a sailor's tolerance of ethanol, and by producing this compound it can poison the well from which its teetotal microbial competitors might otherwise drink. Better even than that, yeast has evolved a gene that enables it to use accumulated ethanol as an energy source, so it can drink from the poisoned well. These biochemical talents equip *S. cerevisiae* supremely well to capture, protect, and then consume the sugars found in fruit.

Saccharomyces cerevisiae is not the only species of yeast able to produce alcohol, but it is uniquely equipped to consume it as an energy source. Alcohol-producing yeasts share a gene for an enzyme called alcohol dehydrogenase (ADH) that is used in ethanol production, but *S. cerevisiae* also has a second, modified copy of the gene that enables it to reverse the process with an enzyme called ADH2. It is ADH2 that makes *S. cerevisiae* such a successful alcoholic.

When did *S. cerevisiae* acquire ADH2? Was this an adaptation to the long association of this yeast species with wine fermentation, perhaps? Was ADH2 the product of yeast domestication? Studies of molecular evolution suggest not, and offer instead an even more intriguing possibility. The origin of the ADH2 gene in the evolutionary history of *S. cerevisiae* has been dated to about eighty million years ago, in the Cretaceous. This is when the flowering plants really got going and fleshy fruit evolved. Of course, fruit are the natural habitat of yeast.

The human species is a very recent player in evolutionary history and we would do well to remember that the bread and the beer that are furnished us by nature have their own evolutionary tales to tell that are mostly much more ancient than our own. The twists and turns of evolution are full of surprises, but a recurring theme can be recognized beneath the anarchy of adaptation and counteradaptation that evolves among species when they interact with one another. We could call it John Barleycorn's theme, because, no matter what the adversity, somewhere a seed of evolutionary success always sprouts from the clay burial ground of defeat.

{ 16 }

Realm of Illusion

COFFEE

The discovery of coffee has enlarged the realm of illusion and given more promise to hope. ISIDORE BOURDON (1796–1861)

The natural history of the coffee bean and the history of the beverage are ineluctably united by a small molecule that the plant uses to defend itself, but which has stimulated a huge appetite in humans. The subversion of caffeine from a defensive role in the natural history of the coffee plant to a motive force that has driven humans to disperse it to every corner of the globe where climate will let it grow is just another twist in the serpentine course of evolution. A path that is always blindly turning corners, reversing fortunes, and springing surprises. If there were purpose in it, evolution would be the supreme practical joker, surreptitiously poisoning one moment and intoxicating the next for its own cruel, cosmic amusement. Instead, evolution is oblivious to any ultimate end. Every invention of evolution is only what we make of it. And what we *do* make of the coffee bean!

Coffee beans are the world's most prized seeds, more valuable in trade than essential foodstuffs like wheat, corn, rice, or soybeans. In fact, coffee is second only to oil in its value in world commerce. An incredible 400 billion cups of coffee are drunk each year. As a ballpark illustration, imagine Yankee Stadium as the world's largest bottomless coffee cup (perhaps with the giant baseball bat that stands outside serving as a cup handle), and you could get 85 refills a year from the global coffee jug before it would run dry.

Try a little caffeinated subversion yourself and ask the local speciality coffee shop for a few unroasted green beans. If they really serve fresh coffee they should have a supply that they roast daily, and if the raw beans are really fresh a few should germinate when you plant them in warm and moist conditions. That's when caffeine begins to fulfill its natural function in the plant. As germination begins, the embryo coffee plant absorbs all the food stored in the endosperm of the seed and with it the caffeine in the bean. Caffeine is an all-purpose defensive compound that is poisonous to insects, inhibits the growth of bacteria and fungi, kills slugs and snails, and even inhibits the growth of plants. The latter effect does not result in the coffee plant poisoning itself, because caffeine is chemically inactivated within the plant's tissues and cloistered out of harm's way.

When the seed germinates, some of the caffeine in the seedling leaks from its roots into the soil, where it may have a protective effect against pathogens and may interfere with the growth of competing plants. The growth of the first, tender leaves of the seedling is crucial to the subsequent survival of the coffee plant. These leaves are heavily protected with caffeine, which is present in their juice at a concentration ten times that found in a cup of espresso coffee. Perhaps Starbucks ought to serve coffee bean sprouts? Mature coffee leaves also contain caffeine, but this is concentrated around the margin of the leaf, where insects are likely to take their first bite.

In the home of coffee in the highlands of Ethiopia the first beverage to be brewed with coffee was probably not made with its seeds, but was a tea made with its leaves. At their best, tea and coffee cannot be confused, but at their worst who hasn't felt sympathy with Abraham Lincoln, who is reputed to have said when offered a suspect beverage, "If this is coffee, please bring me some tea; but if this is tea, please bring me some coffee."

A pair of coffee beans form the pit of a red or yellow, cherry-like fruit that grows on a small evergreen tree, *Coffea arabica*. This species is the original source of the beverage and the one from which quality coffee is made. *C. arabica* beans contain less than 1.5 percent caffeine. A cheaper, inferior product is made from the beans of the related species *Coffea canephora* that originates from Zaire. One connoisseur and writer on coffee has described the *Robusta* variety of coffee made from the very widely grown *C. canephora* as "*Arabica* coffee's crude, boorish, sour, uncivilized black-hearted cousin." *Robusta* is more disease resistant and its seeds have a higher caffeine content than *Arabica*.

In cultivation, the coffee tree is clipped and grown as a bush. Coffee beans ripen slowly on the bush, which at any one time bears both flowers and fruit at different stages of ripeness. Unripe beans contain chemicals that can ruin the taste of a whole batch when roasted, so coffee is harvested by handpicking cherries when each has reached the point of ripeness; otherwise, unripe cherries must be removed from the crop if it has been indiscriminately harvested by machine. After the fruit pulp has been removed and the beans have been dried they are light green in color, raw and pregnant with undeveloped flavors. Roasting at a temperature of about 220 degrees Celsius performs molecular magic on the flavor of the raw bean, increasing the chemical complexity of its aroma.

A roasting bean is a miniature pressure cooker containing thousands of cells heated to bursting point. As the water in the cells vaporizes it puffs up the bean to twice its original size, carbohydrates break down to simple sugars, and these caramelize, giving the bean its rich brown hue. Under the heat and pressure, other chemical reactions take place, releasing more than eight hundred different molecules that contribute to the delicious aroma of the roast. The mixture is so complex that no good artificial coffee flavor has ever been successfully manufactured. Copious amounts of carbon dioxide are generated by all these reactions (up to twelve liters by a kilo of beans) and some of this remains trapped inside the cells of the bean, ready to create the foam, or *crema*, on top of your espresso when expelled under pressure from the ground beans by the espresso machine.

Caffeine is chemically unaffected by roasting. The alkaloid is nearly flavorless and odorless, which is why decaffeinated coffee can still taste good. All the flavor and aroma come from the coffee oils, which constitute only 3 percent of the roasted bean. Raw coffee beans can be stored safely without deterioration, but the moment the oils are released by roasting they are susceptible to oxidation and decay. The full aroma and flavor of coffee is an evanescent thing that must be captured as near the moment of roasting as possible. Vacuum packing or freezing roasted beans helps arrest the decay of flavor, but the tradition in Ethiopia, the homeland of coffee, is to roast, grind, and prepare coffee all in one ceremony so that drinkers can enjoy the aroma as well as the taste and nothing is lost in the process. A full Ethiopian coffee ceremony is a social occasion that can last two hours.

Once the Ethiopians had discovered coffee, at some uncertain date, but perhaps two thousand years ago, caffeine began to influence social evolution, and this story, as full of twists and turns as any biological history, became intertwined with the evolution of the plant. Indeed, the path by which coffee spread from Ethiopia through the Arab world, to Europe and its Asian colonies and then to the Americas, need no longer be the subject of historical speculation. The history of coffee's spread has now been reconstructed by tracing the genealogy of coffee varieties recorded in their DNA.

Coffee was first exported around the Muslim world from the port of Mocha in Yemen, where plants from Ethiopia were grown and traded. The coffee plants grown in Yemen today display only a fraction of the genetic diversity found in coffee in the Ethiopian highlands, indicating that Yemen's plants probably descend from only a few individuals, maybe a handful or two of beans, that were carried there from Ethiopia. When this happened is not known, but it may have been in the sixth century AD when the Ethiopians invaded Yemen and ruled there for a brief period. Coffee in Yemen was adopted by Arab Sufi monks who used the drink to help them stay awake for midnight prayers. Pilgrims to the holy places of Islam spread the drinking of coffee throughout the Muslim world, so that it had become a valuable trade commodity by the end of the fifteenth century.

Arab coffee houses began the association of coffee with independent thought and political dissent. In 1511, Khair-Beg, the governor of Mecca, decided he would close the coffee houses of the city because they were the source of satirical verses about him. The ban was soon reversed by the sultan in Cairo, who was a coffee drinker, but as coffee spread so did its reputation among rulers as a stimulant to revolt. The most notorious persecutor of coffee drinkers was the grand vizier Kuprili of Constantinople, who had repeat offenders sewn into leather bags and thrown into the Bosphorus.

Since its appearance in Europe in the sixteenth century, coffee has been variously prescribed for its healthful effects and proscribed for its evil ones. An anonymous broadside published in England in 1674 called "Women's petition against Coffee" proclaimed that "Never did Men wear *greater Breeches*, or carry *less* in them of any *Mettle* whatsoever. . . . They come from it with nothing *moist* but their snotty Noses, nothing *stiffe* but their joints, nor *standing* but their ears." The reference in this roasting of coffee drinkers to their lack of moisture derives from the fear expressed by certain medical men that the diuretic effects of coffee would cause addicts of the drink to pee themselves to death.

The broadside was headed "The Humble Petition and Address of Several Thousand Buxome Good-Women, Languishing in Extremety of Want . . . ," but sounds to me more like a male fantasy dreamed up in a tavern. Alehouse keepers were distinctly worried that the new coffee houses would deprive them of trade. The true motives of the anonymous authors of the petition are let slip at the end where "we pray that drinking COFFEE be forbidden to all Persons under the Age of Threescore and that Lusty Nappy Beer and Cock Ale be Recommended to General Use." Even a century later, Frederick the Great of Prussia favored beer over coffee drinking, which he thought was sapping the fertility and health of the nation. In 1777 he issued an edict stating, "It is disgusting to notice the increase in the quantity of coffee used by my subjects, and the amount of money that goes out of the country as a consequence. . . . My people must drink beer."

The contrary point of view scarcely needs rehearsal. Coffee has inspired artists, scientists, and politicians and made many fortunes. The discovery of coffee not only enlarged the realm of illusion, as Isidore Bourdon poetically put it, but has also enlarged the realms of freedom, culture, commerce, and reason, all of which give more promise to hope. Paul Erdös, one of the greatest mathematicians of the twentieth century, famously said "A mathematician is a device for turning coffee into theorems." One can only imagine how much coffee he must have consumed, since he published more articles than any other mathematician, ever. The composer Ludwig van Beethoven was a coffee drinker who used precisely sixty beans per cup to make his own brew (remarkably near to the fifty to fifty-five beans required to make a modern cup of espresso). Johann Sebastian Bach composed a coffee cantata, setting to music the refrain "How sweet is the taste of coffee, more charming than a thousand kisses, softer than Muscatel wine." (It is worth remembering that this is how good coffee ought to taste, though it very rarely does in my experience.) The French novelist Honoré de Balzac drank sixty cups a day (which must have provided a near-lethal dose of caffeine) to fuel his prodigious literary output, saying "As soon as coffee is in your stomach, there is a general commotion. Ideas begin to move . . . similes arise, the paper is covered. Coffee is your ally and writing ceases to be a struggle."

William Harvey, seventeenth-century physician and discoverer of the circulation of the blood, was ahead of his time in his appreciation of coffee as well as in his science. He died in 1657, before coffee became widely used in England, but he bequeathed fifty-six pounds of beans to his colleagues in the London College of Physicians, averring that "this little fruit is the source of happiness and wit!" and requesting them to commemorate his death by meeting together monthly to drink coffee until the supply of beans was exhausted.

Modern science confirms many of the benefits of moderate coffee consumption that were claimed for it by users in earlier centuries. Caffeine blocks the action of a substance called adenosine, which is a widespread regulator of the function of the central nervous system. Adenosine acts like a brake on the firing of neurons, so when caffeine gets in the way of this brake, the human machine speeds up. Adenosine accumulates during waking hours, eventually inducing sleep. Caffeine prevents adenosine from performing this function, which is why coffee keeps you awake. In the succinct language of science, a review of the effects of caffeine states that a low to moderate dose "increases arousal, vigilance, and motor activity; decreases the need to sleep; produces sensations of well-being and energy; and facilitates cognitive capacities." To which conclusion William Harvey would undoubtedly have raised his cup.

As it is now, the social aspect of coffee drinking was highly important in the eighteenth century and the seeds of many important commercial and scientific institutions were germinated in coffee houses. The shipping underwriters Lloyd's of London began in a coffee house of that name; the London Stock Exchange started in Jonathan's Coffee House in Exchange Alley; and in Boston the first public auction of stock took place at the Merchant Coffee House. The New York stock exchange began at the Tontine Coffee House on Wall Street. The East India Company, through whose exploits India became the jewel in the British imperial crown, had its unofficial headquarters in the Jerusalem Coffee House in London; the Royal Society of London was founded in Tillyard's coffee house.

In Paris, the French Revolution was plotted in coffee houses, and on the other side of the Atlantic, a Boston coffee house and tavern called the Green Dragon became the headquarters of the American Revolution. English loyalists had many meeting places in Boston to choose from, including the Crown Coffee House, the King's Head, and the British. When the English were defeated, "The British" immediately changed its name to "The American." For revolutionary Americans, coffee was an alternative to the overtaxed tea that the British East India Company shipped to Boston. A rallying song for the Boston Tea Party went:

Rally, Mohawks—bring out your axes!
And tell King George we'll pay no taxes
on his foreign tea!
His threats are vain—and vain to think
To force our girls and wives to drink
his vile Bohea!
Then rally boys, and hasten on
To meet our Chiefs at the Green Dragon.

The boycott of tea (of which Bohea was a variety) was short-lived, and the East India Company was back in business supplying tea to America by the early 1800s. The power of commerce is rarely suppressed for long. Indeed, commerce acquired complete control of the evolution of coffee. The genetic diversity of the plant, already diminished by its transfer from the Ethiopian highlands to Yemen, was reduced still further when plants were smuggled out of Yemen and used to establish plantations in Java in Indonesia in 1690 and on the island of La Réunion in the Indian Ocean soon afterward. A single plant taken from Java was cultivated in the botanical garden at Amsterdam in the Netherlands and from this one plant descend all the millions of trees of the Typica variety grown around the world. The other main variety of *Coffea arabica* is Bourbon from La Réunion. Bourbon is not quite as genetically depauperate as Typica, but the very low genetic diversity in plantations of *Coffea arabica* must be an important factor in their vulnerability to diseases such as the leaf rust that destroyed Brazil's coffee crop in 1970. The response by growers was to plant "*Arabica* coffee's crude, boorish, sour, uncivilized black-hearted cousin," *Coffea canephora*, which is more disease resistant.

An ironic twist was given to the intertwined relationship between the biological and cultural evolution of coffee when, at the end of the nineteenth century, a German named Ludwig Roselius realized that there might be a market for coffee without caffeine. Roselius's father was a coffee taster by profession, and Ludwig ascribed his father's early death to caffeine-poisoning. Ludwig Roselius invented a process by which caffeine could be extracted from raw coffee beans without impairing the normal aroma of coffee that develops with roasting. About 10 percent of the market for coffee is for decaffeinated beans. In 2004 coffee breeders discovered a coffee plant in Ethiopia that produces beans that are naturally lacking in caffeine, from which they intend to breed a new variety. With this development the influence of coffee on culture will have come full circle. The demand for the aroma, the taste, and the companionship engendered by the seed will have selected and propagated coffee trees disarmed of the very molecule that gave impetus to the rise of coffee culture.

{ 17 }

Nourishment & Inspiration

GASTRONOMY

In fact, seeds gave early humans both the nourishment and the inspiration to begin to shape the natural world to their own needs. Ten thousand turbulent years of civilization have unfolded from the seed's pale repose.
HAROLD MCGEE, *On Food and Cooking*

A little knowledge of seeds can be a delicious thing. Tired of haute cuisine? Try oat cuisine instead. Know the properties of your seed ingredients and you can bake the tastiest wheat

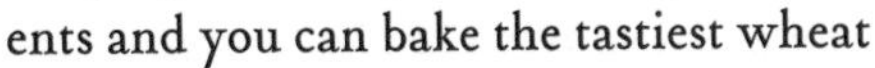

bread and cornbread, concoct nutritious salads, cook perfect rice, beans, and lentils, and figure out why popcorn pops and alfalfa sprouts. Seeds are designed by evolution to store food for juvenile nourishment. This simple fact is the key to all their properties as food for humans. It is why they can be kept so easily without going bad, why they are packed with energy stored as starch or fat; why they often have high protein content, and also why they are so often guarded with poisonous or repellent substances. The science of cooking with seeds is a grand campaign based on the wily subversion of the gifts a mother plant makes to her offspring. It is intelligent plundering and piracy.

I emphasize *intelligent*. There are plenty of nutty claims for the nutrient effects of seeds out there. One of my favorites is the claim made for the graham cracker by its inventor Sylvester Graham, who believed that its wholegrain recipe would help weaken sexual desire among followers of his cult. Truly crackers, but harmless. Not so harmless was the "macrobiotic" diet that became a cult in 1960s North America. Adherents followed a regime of ten steps that increasingly restricted their diet until the final stage, when only brown rice, salt, and herbal teas were allowed. Few reached this ultimate stage, but some of those who did died of malnutrition.

As with many food fads, there is a kernel of truth in the efficacy of diets based on whole grains. Extensive medical studies have shown that people who include significant amounts of whole grains in their diets have a substantially lower risk of cardiovascular disease. Dr. Sylvester Graham would no doubt have been mortified to learn that the cardiovascular benefits of whole grains mean that you can have more sex, not less. As part of a balanced diet, eating more whole-grain foods such as wild rice, popcorn, oats, brown rice, barley, and wholemeal bread is good for you.

Let us explore oat cuisine. In the first dictionary of the English language, published in 1755, its compiler, Dr. Samuel Johnson, defined oats as "A grain, which in England is generally given to horses, but in Scotland appears to support the people." Since Dr. Johnson also notoriously said that "the noblest prospect which a Scotchman ever sees is the high road that leads him to England!" his opinions, like porridge, ought to be taken with a pinch of salt. From a practical point of view the oat plant *Avena sativa* has the virtues that it thrives in the damp climate of northwest Europe and its seeds have a high nutritive value, containing about 15 percent protein and 8 percent fat. As the Scots poet Robert (Rabbie) Burns put it, oatmeal porridge is "Chief of Scotia's food." The rural poverty that forced many Highlanders in the days of Rabbie Burns to subsist on watery porridge is thankfully gone. Now, we can appreciate the culinary and nutritional qualities of oats without the influence of penury or prejudice.

Oats are a good source of soluble dietary fiber, which has many health benefits, including lowering the risks of coronary heart disease and the most common form of diabetes (type 2). The soluble dietary fiber gives oatmeal porridge the smooth, thick consistency that connoisseurs look for. It is also used in baking to soften and moisten bread and can be used as a thickening agent in soups and stews. These properties of oats derive from water-absorbing, indigestible carbohydrates called beta-glucans that are concentrated in a layer just beneath the skin (called the aleurone) that surrounds the oat seed. The natural function of beta-glucans is to take up and store water when the seed germinates. We pirate beta-glucans from oats to use this water-holding property in our food and cooking.

Each year contestants gather at the World Porridge-Making Championships at the village of Carrbridge, in Inverness, Scotland, to compete for the Golden Spurtle (a porridge stirrer). But there is really no mystery to making good porridge. It can be made with just oatmeal and water on the hob or in a microwave, though a creamier version requires milk. The real challenge comes in using plain oatmeal porridge to create something different. Heston Blumenthal, the chef-proprietor of the Fat Duck, rated the world's best restaurant in 2005 and with three Michelin stars, invented snail porridge, which has become one of his signature dishes. The full recipe is available on the Internet, but the principle is to use porridge oats in place of rice in what might otherwise be called a snail risotto. The porridge is made with stock that has been used to cook the snails. A garlic butter made with mushrooms, shallots, Dijon mustard, ground almonds, Parma ham, and parsley is beaten into the porridge and the cooked snails are either chopped and mixed into the porridge or placed on top with layers of shredded Parma ham and finely sliced fennel dressed with vinegar and walnut oil. No doubt Dr. Samuel Johnson would have had something colorful to say about snail porridge, but I have tried it and the dish deserves several golden spurtles.

Though seeds, which evolved to nourish juvenile plants, provide such good food for animals, including ourselves, they are rarely able to satisfy every dietary requirement. We get the most out of seeds when different species are combined in the diet. A diet of just corn or beans would be a deficient one, but when combined they complement each other in a nutritional partnership that fostered and sustained for millennia the cultures of Mesoamerica that first domesticated them.

Cereals such as wheat, rice, and corn contain all eight essential amino acids, but have lower amounts of two of them, lysine and threonine, than are required in the human diet. Pulses such as soybeans, peas, or lentils contain adequate levels of these amino acids, but are deficient in two others: cysteine and methionine. Thus, a diet or a meal that combines a cereal with a pulse is balanced by the complementary amino acid composition of the two kinds of seeds. Many traditional cuisines have their own ways of combining cereals and pulses, enabling poor people to get by with little or no meat: beans and tortillas in Mexico; rice and peas in the Caribbean; hummus (made from chickpeas and sesame seed) and pita bread in the Middle East; dal (lentils) and rice in India; chickpeas and couscous (made from wheat) in North Africa, rice and tofu (bean curd) in China and Japan.

We need to be cautious, though, before assuming that foods that work in one culture will necessarily be adequate in another. Vitamin B_{12} deficiency is a risk for vegans, who consume no meat, and it can lead to serious health problems. In the mid-1970s it was found that orthodox Hindus who had been quite healthy on a vegan diet in their native India began to suffer from a high incidence of megaloblastic anemia after living for some time in England consuming the same diet. The cause was traced to vitamin B_{12} deficiency, which in India was prevented by insect contamination of grains. The same foodstuffs purchased in England were uncontaminated and therefore contained no animal source of vitamin B_{12}. In fact, neither plants nor animals can manufacture B_{12}, so all animals obtain their supply of this essential vitamin either directly or indirectly from bacteria, which are the only organisms able to make it.

The reason why a seed that suffices to nourish a juvenile plant is rarely enough for any animal is that plants are nutritionally much more self-sufficient than animals. Given the basic starting materials and an energy supply from a seed or the sun, plants can make all of the complex molecules that they need, including the twenty amino acids that are used to build proteins. Seedlings can make for themselves what they do not acquire in their starter kit from mother. We humans, on the other hand, can manufacture only twelve of the twenty amino acids. The twelve amino acids that our cells can make they produce from the eight amino acids that they cannot, so we must get those eight essential amino acids from our food. Meat of almost any kind can supply them, but there is one seed that also contains the full complement of eight. It is the seed of *Chenopodium quinoa*, a plant in the spinach family.

Quinoa (pronounced "keen-wa") was domesticated in the Andes about seven thousand years ago and was a staple of the Incas. Such perfect food requires protection from animals, so it should be no surprise that many varieties of quinoa seed are defended by bitter-tasting compounds in their outer skin. These compounds can be removed by washing and rubbing in cold water. Once rendered defenseless, quinoa's tiny seeds can be cooked like more familiar grains such as rice, fried and added to salads or soups, or popped to make doll-proportioned popcorn.

The nutrient qualities of our food are wedded to their taste by an evolutionary relationship that runs so deep that it has shaped the physiology and the psychology of how we perceive flavor. Five types of taste receptor on the tongue tell us whether a food is potentially good to eat or may be poisonous. Two distinct receptors detect sour and bitter substances that are associated with rotten or poisonous food, while receptors for saltiness, sweetness, and umami flavors indicate nutritious foods. Umami is the meaty flavor of certain amino acids and indicates food that contains protein. Monosodium glutamate (MSG) stimulates umami receptors very strongly, which is why it is used as a flavor enhancer. MSG occurs naturally in fermented soybean paste, which is the source of the miso and soy sauce used to flavor Chinese and Japanese dishes. However, there is more to flavor than just the five basic tastes that stimulate specific receptors on the tongue.

Flavor sensations are a complicated confection created in the brain from the combined inputs of all five senses. Just as the three color receptors in the human retina allow us to discriminate thousands of colors, so our sense of taste has a far larger repertoire than just saltiness, sourness, sweetness, bitterness, and umami. In addition, there are hundreds of receptors for different odors that can be detected in the nose. If you have ever noticed how different and bland food tastes when your nose is blocked by a cold, you will already appreciate the important role that the sense of smell plays in the perception of flavor. As an experiment, try tasting some food while pinching your nose. Without a flow of air through the nostrils, the aroma of the food you have in your mouth cannot be drawn over the upper lining of your nose where the cells that sense odors are located. But even with your nose pinched you should still be able to taste a lick of salt, because this stimulates one of the five receptor types in your tongue. Although the other four basic tastes should also be detectable in this way, other flavors, like vanilla for example, will not be.

The color, the feel of food in the mouth, and even the sound it makes as it is chewed all influence flavor perception. Mouth feel is particularly important in the pleasurable sensation derived from eating chocolate. The fat (cocoa butter) from the cocoa beans that are used to make chocolate gives it the velvety texture that we so enjoy. This texture may actually be more important in producing chocolate cravings in the chocoholic than the pharmacological effect of theobromine, an alkaloid (like caffeine) that is also present in chocolate. There is a simple way to test this for yourself. Drinking chocolate, or cocoa powder, contains all the pharmacological constituents and sugar found in a bar of chocolate, but without the cocoa butter. If you have a craving for chocolate, you can see whether a cup of cocoa will satisfy you. You can also try eating white chocolate, which has the sugar and cocoa butter but not the pharmacological compounds found in normal chocolate. Experiments of this kind have found that white chocolate satisfies more than cocoa, but that (not surprisingly) brown chocolate is rated tops.

Because our perception of the world involves the brain interpreting inputs from receptors of various kinds, manipulation of the receptors can create illusions. For example, berries of the West African tree *Synsepalum dulcificum* contain a protein that interferes with taste receptors in the tongue and causes sour foods to taste sweet. There was an extensive research program in the 1970s to try to exploit the "miraculous" berry of *S. dulcificum* as an artificial, low-calorie sweetener, but it failed because the protein responsible for the effect is unstable. Only fresh berries work. Presumably the miracle berry tricks birds into believing that they are feeding on sweet, nutritious fruit. They will then carry away the miracle berry's seeds without the tree having to pay the customary fare in expensive sugar. Actually, it turns out that the sweet sensors on the tongue are quite easily fooled. The artificial sweeteners saccharine and aspartame do it, and even adding salt to a pineapple can make it taste sweeter.

A rare visual food illusion is created by the unique optical properties of pumpkin seed oil, which is a speciality of Austria and Hungary, where it is used as a salad dressing. The color of the oil appears to change from bright red in the bottle to emerald green on a plate or when mixed with yoghurt, but this is an illusion and is not caused by any chemical change in the oil. The explanation for the change lies in the retina of the human eye and in the peculiar spectral properties of the oil. Recall from chapter 9 that our color perception depends on three receptors, one with a peak at short wavelengths (blue), one at mid wavelengths (green) and one at long wavelengths (red). Pumpkin seed oil has a narrow window in its spectrum that lets through green light and a broad window that lets through red. A thin layer of the oil lets through enough green light to stimulate the green receptors more strongly than the red ones, but if the layer of oil is more than 0.7 millimeter thick, it transmits relatively less green light and the red receptors are more strongly stimulated. A layer precisely 0.7 millimeter thick lets through light that stimulates green and red receptors equally and the oil appears yellow.

A "grasshopper mind" is the only way to describe the human brain's interest in odors. When an odor first appears it evokes a strong reaction, but most smells quickly become unnoticeable. Smells disappear even without any change in the actual concentration of odor molecules. Most people have probably noticed this phenomenon, but how many have considered why it happens? A new smell sends an alert to the brain to which we can react in an appropriate manner, but if we take no action that alters our exposure to the smell, the brain acts as though the information contained in the odor is redundant and the smell is ignored. This makes great sense from a survival point of view because it enables us to remain alert to new threats or new opportunities, without being distracted by irrelevant details in our surroundings.

However, from the perspective of the chef who wants to grab his or her clientele by the taste buds, the tendency for taste and odor receptors to tire so easily must be like trying to coach a team of flabby geriatrics to play major-league baseball. One solution is to create concentrated jabs of flavor by encapsulating foods so that they burst upon the palate and continually arouse the obdurate olfactory system. Heston Blumenthal at the Fat Duck uses small cubes of jelly to encapsulate flavors. Chocolate chips in cookies, dried fruit pieces in cakes, and whole grains in bread all use the same principle to deliver a kick to the taste buds.

Seeds are natural capsules of flavor, but they usually need to be roasted for their flavorful potential to be realized. As we saw in chapter 16, roasting greatly increases the number of flavor compounds in coffee and produces the heavenly aroma. Roasting has the same beneficial effects on the aroma and flavor of other seeds, including spices such as cumin and coriander, and peanuts, sunflower seeds, pumpkin seeds, chestnuts, almonds, and many more. The release of flavor and aroma is caused by the physical effect of roasting, which ruptures cells, allowing fragrant oils to escape, and through a collection of chemical transformations that take place at high temperature. Chief among these is the Maillard reaction, which produces a cornucopia of aroma compounds by joining sugars and amino acids together. Particular versions of this reaction produce the molecules responsible for the distinctive toasty aroma of baking bread, the nutty odor of roasting peanuts, the smell of popcorn and other evocative kitchen delights like the smell of french fries. The chemical nature of this reaction, which is also responsible for browning in meat and other roasted foods, was published by the French chemist Louis Camille Maillard in 1912. In just seventy-seven lines that included the title of his paper, Maillard described the mechanism of the reaction between sugars and amino acids, the methods he had used in his experiments, and all the implications of the discovery, including its possible role, now well established, in pathological disorders brought on by diabetes. Unfortunately, his work was so far ahead of its time that it was ignored for over twenty years, by which time Maillard had become severely disabled by a typhoid infection he had caught while working on the medical control of the disease during the First World War. But his name lives on in the way that aromas linger, renewed by every fresh discovery of new instances of the Maillard reaction.

Not only the flavor, but the texture of seeds may be radically altered by roasting them. In the winter of 1842 Henry David Thoreau, New England backwoodsman, botanist, philosopher, poet, and believer in the power of seeds, recorded in his journal: "I have been popping corn to-night, which is only a more rapid blossoming of the seed under a greater than July heat. The popped corn is a perfect winter flower, hinting of anemones and houstonias.... By my warm hearth sprang these cerealious blossoms; here was the bank where they grew."

There may be more significance in his cerealious blossoms than even Thoreau guessed, for there is archaeological evidence that popping kernels in the embers of a fire may have been the earliest method by which corn was first cooked in Mexico. Astonishingly, some corn kernels recovered from deposits that are thousands of years old can still be popped. So the first popcorn blossoms marked one of the most momentous springs in all of human history—the earliest shoots of New World agriculture.

Many cereal grains can be puffed or popped, but the champion of endosperm explosiveness is *Zea mays everta*. This corn variety has small, herd kernels bound in a dense straightjacket of cellulose fibers that transmit heat efficiently from the outside of the seed to the tiny pressure cooker within. As the interior of the kernel heats up, the moisture in it vaporizes and the endosperm softens. The pressure inside the kernel builds up to seven times that of the atmosphere before the straightjacket bursts. The sudden release of pressure causes the softened endosperm to explode, puffing up and then stiffening as it cools. If you want to make successful popcorn on a hob, make sure that the lid does not seal the pan. If it does, the pressure difference that drives the explosion of the endosperm will be less and the popcorn will be tough and chewy rather than light and fluffy.

IN EVERY CHAPTER of this book we have encountered examples of ways in which evolution finds new uses for old devices and continually subverts the stratagem of one organism to the advantage of another. Truly, science can be stranger than fiction. Isn't there the plot of a gothic, romantic novel in the story of how the pollen tube that delivers sperm to its tryst with the ovule in the bridal bed of the flower evolved from the implement by which the parasitic male siphoned food from the unfertilized ovule? What strange twist of fate fed the diseased imagination of witch hunters in Salem with ergot-infested rye that turned the bread of holy communion blood red? Perhaps the ancient Greek dramatist Aeschylus would have recognized the vengeance of Zeus in the fate of the sad cypress of the Sahara as it treads the evolutionary road to extinction? Surely divine justice for a reproductive system in which pollen usurps the genetic birthright of the seed.

But other cheats do prosper, like the interloping insects that take a free ride on the intricate pollination system of the yuccas or figs, or the transposons that selfishly multiply in the genomes of corn and human alike. Recall the battle for control of germination between the acorn and the squirrel or the way that yeasts poison fermenting barley seeds with alcohol, denying their nutritious contents to other microorganisms. For thousands of years we have subverted that particular yeast stratagem for our own enjoyment. Cooking is evolutionary subversion too. When you enjoy seeds, you can enrich the experience by sparing a thought for their fascinating evolutionary journey to your plate.

Scientific Names

Plants, animals, and microorganisms mentioned in the text

COMMON NAME	SCIENTIFIC NAME
—	*Alsomitra macrocarpa*
African oil palm	*Elaeis guineensis*
Ash	*Fraxinus* spp.
Barley	*Hordeum vulgare*
Blackberry	*Rubus* spp.
Black-capped chickadee	*Parus atricapillus*
Black-legged tick	*Ixodes scapularis*
Brazilian zebra wood	*Centrolobium robustum*
Brewer's yeast	*Saccharomyces cerevisiae*
Bunt fungus	*Tilletia tritici*
Cacao	*Theobroma cacao*
Candlenut palm	*Aleurites molucanna*
Coffee (Arabica)	*Coffea arabica*
Coffee (Robusta)	*Coffea canephora*
Corn (maize)	*Zea mays*

COMMON NAME	SCIENTIFIC NAME
Daffodil	*Narcissus pseudonarcissus*
Date palm	*Phoenix dactylifera*
Dipterocarp trees	family Dipterocarpaceae
Double coconut	*Lodoicea maldivica*
Douglas fir	*Pseudotsuga menziesii*
Ergot fungus	*Claviceps purpurea*
Fig	*Ficus* spp.
Giant redwood	*Sequoiadendron giganteum*
Globeflower	*Trollius* spp.
Gray squirrel	*Sciurus carolinensis*
Gypsy moth	*Lymantria dispar*
Hawthorn	*Crataegus* spp.
Huitlacoche or Mexican corn smut	*Ustilago maydis*
Kidney vetch	*Anthyllis vulneraria*
Laurel	*Daphne odora*
Liverwort	—
Lodgepole pine	*Pinus contorta*
London plane	*Platanus X hispanica*
Lyme disease spirochete	*Borellia burgdorferi*
Maidenhair tree	*Ginkgo biloba*
Maple	*Acer* spp.
Miracle berry	*Synsepalum dulcificum*
Neem tree	*Azadirachta indica*
Oak	*Quercus* spp.
Oat	*Avena nativa*
Orangutan	*Pongo abelii*
Pecan	*Carya illinoensis*
Pecan nut casebearer	*Acrobasis nuxvorella*
Pecan weevil	*Curculio caryae*
Phylloxera grape aphid	*Daktulosphaira vitifoliae*

COMMON NAME	SCIENTIFIC NAME
Pine siskin	*Carduelis pinus*
Pine squirrel	*Tamiascurus hudsonensis*
Primrose	*Primula vulgaris*
Quinoa	*Chenopodium quinoa*
Red crossbill	*Loxia curvirostra*
Rye	*Secale cereale*
Sahara cypress	*Cupressus dupreziana*
Seagrasses	Marine flowering plants belonging to the four families *Posidoniaceae*, *Zosteraceae*, *Hydrocharitaceae*, and *Cymodoceaceae*.
Senita cactus	*Lophocereus schottii*
Stick insects	Insects of the order Phasmatodea
Sunflower	*Helianthus annuus*
Swamp loosestrife	*Decodon verticillatus*
Violet	*Viola* spp.
Wall barley	*Hordeum murinum*
Whitebeam	*Sorbus* spp.
White-footed mouse	*Peromyscus leucopus*
White-tailed deer	*Odocoileus virginianus*

Sources and Further Reading

Chapter One } AN ORCHARD INVISIBLE: SEEDS

SOURCES

A seed hidden in the heart of an apple: B. E. Juniper and D. J. Mabberley, *The Story of the Apple* (Portland: Timber Press, 2006), 139.

I have great faith in a seed: H. D. Thoreau, *Faith in a Seed* (Washington, DC: Island Press, 1993).

To see things in the seed: http://en.wikiquote.org/wiki/Laozi (accessed 18 February 2007).

FURTHER READING

Seeds: Time Capsules of Life, by R. Kessler and W. Stuppy (London: Papadakis, 2006), is a work of science hidden inside a magnificent coffee table book with stunning color pictures of seeds on every page. If you have any questions about the science of seeds, the first port of call should be your library's copy of *The Encyclopedia of Seeds: Science, Technology, and Uses*, ed. M. Black, J. Bewley, and P. Halmer (Cambridge, MA: CABI, 2006). This book is comprehensive on its subject, if a little weak for my taste on evolutionary aspects of its subject.

Chapter Two } FIRST FORMS MINUTE: EVOLUTION

SOURCES

Land plants "are made from a sea recipe": E. H. Corner, *The Life of Plants* (Chicago: University of Chicago Press, 1964), 161.

All land plants . . . descend from a single ancestor: K. J. Willis and J. C. McElwain, *The Evolution of Plants* (Oxford: Oxford University Press, 2002).

A list would cover many pages: E. H. Corner, *The Life of Plants* (Chicago: University of Chicago Press, 1964), 104.

There should be no monotony / In studying your botany: excerpted from "Botany" by Berton Braley, *Plant Physiology Information Website*, http://plantphys.info/botany.poem.html (accessed 8 March 2007).

Fern seed could be collected at midnight on Midsummer Night's eve: R. C. Moran, *A Natural History of Ferns* (Portland, OR: Timber Press, 2004).

We have the receipt of fern-seed: Shakespeare, *King Henry IV, Part One*, in *Complete Works of William Shakespeare* (London: Collins Classics), act 2, scene1.

Homeopaths still believe in the power of remedies: E. Ernst, "Is Homeopathy a Clinically Valuable Approach?" *Trends in Pharmacological Sciences* 26 (2005): 547–48.

One specimen . . . even survived the blast of the atomic bomb: http://www.xs4all.nl/~kwanten/hiroshima.htm (accessed 2 March 2007).

It takes up to four months before her egg is ready for fertilization: Y. Nakao, K. Kawase, S. Shiozaki, T. Ogata, and S. Horiuchi, "The Growth of Pollen and Female Reproductive Organs of *Ginkgo* between Pollination and Fertilization," *Journal of the Japanese Society for Horticultural Science* 70 (2001): 21–27.

In 1995 when double fertilization was discovered in *Ephedra*: W. E. Friedman, "Organismal Duplication, Inclusive Fitness Theory, and Altruism—Understanding the Evolution of Endosperm and the Angiosperm Reproductive Syndrome," *Proceedings of the National Academy of Sciences of the United States of America* 92 (1995): 3913–17.

In 1999 new DNA data: W. E. Friedman, "Comparative Embryology of Basal Angiosperms," *Current Opinion in Plant Biology* 4 (1999):14–20.

A division of labor between carers . . . and bearers: R. Dawkins, *The Selfish Gene* (Oxford: Oxford University Press, 1976).

Charles Darwin himself saw the difficulty: C. Darwin, *The Origin of Species by Means of Natural Selection* (London: John Murray, 1859), chap. 7.

One of Darwin's chief difficulties has become one of the crowning achievements of his theory: J. E. Strassmann and D. C. Queller, "Insect Societies as Divided Organisms: The Complexities of Purpose and Cross-Purpose," *Proceedings of the National Academy of Sciences of the United States of America* 104 (2007): 8619–26.

The difference in relatedness between parents and seeds therefore produces a conflict of interest: D. Haig and M. Westoby, "Parent-Specific Gene-expression and the Triploid Endosperm," *American Naturalist* 134 (1989): 147–55.

Why 2m:1p endosperm was favored by natural selection: J. A. Stewart-Cox, N. F. Britton, and M. Mogie, "Endosperm Triploidy Has a Selective Advantage during Ongoing Parental Conflict by Imprinting," *Proceedings of the Royal Society of London Series B-Biological Sciences* 271 (2004): 1737–43.

When the relative doses of maternal:paternal genes is manipulated: B.-Y. Lin, "Association of Endosperm Reduction with Parental Imprinting in Maize," *Genetics* 100 (1982): 475–86.

A gene on one particular chromosome that is required for the production of normal endosperm: M. Gehring, Y. Choi, and R. L. Fischer, "Imprinting and Seed Development," *Plant Cell* 16 (2004): S203–13.

FURTHER READING

I recommend two books, though unfortunately neither is written for a popular audience: K. J. Willis and J. C. McElwain, *The Evolution of Plants* (Oxford: Oxford University Press, 2002); and K. J. Niklas, *The Evolutionary Biology of Plants* (Chicago: University of Chicago Press, 1997).

Chapter Three } EVEN BEANS DO IT: SEX

SOURCES

The big evolutionary breakthrough that involved two individuals exchanging DNA occurred very early in the history of life: M. A. Ramesh, S.-B. Malik, and J. M. Logsdon Jr., "A Phylogenomic Inventory of Meiotic Genes: Evidence for Sex in *Giardia* and an Early Eukaryotic Origin of Meiosis," *Current Biology* 15 (2005): 185–91; J. P. Xu, "The Prevalence and Evolution of Sex in Microorganisms," *Genome* 47 (2004): 775–80.

The Greek philosopher Theophrastus: M. Negbi, "Male and Female in Theophrastus's Botanical Works," *Journal of the History of Biology* 28 (1995): 317–32.

The Roman poet Ovid: A. Bristow, *The Sex Life of Plants* (New York: Holt, Reinhart and Winston, 1978).

Daffodils (*Narcissus* species), have a fascinatingly varied sex life: S. W. Graham and S. C. H. Barrett, "Phylogenetic Reconstruction of the Evolution of Stylar Polymorphisms in *Narcissus* (Amaryllidaceae)," *American Journal of Botany* 91 (2004): 1007–21.

"Nor drugs nor herbs will what you fancy prove": Quoted in K. F. Kiple and K. C. Ornelas, eds., *The Cambridge World History of Food* (Cambridge: University of Cambridge Press, 2000), 2:1530.

In 1683 Grew referred to "the foetus or true seed": "Review: The Anatomy of Plants: With an Idea of a Philosophical History of Plants; and Several Other Lectures, Read before the Royal Society by Nehemiah Grew M. D. Fellow of the Royal Society, and of the College of Physitians by Nehemiah Grew," *Philosophical Transactions of the Royal Society* 13 (1683): 303–10.

James Logan, Chief Justice and President of the Council of Pennsylvania: J. Logan, "Some Experiments concerning the Impregnation of the Seeds of Plants," *Philosophical Transactions of the Royal Society* 39 (1735): 192–95.

"I am not born a poet but somewhat of a botanist": Quoted in N. Gourlie, *The Prince of Botanists: Carl Linnaeus* (London: H. F. and G. Witherby, 1953), 28.

"The petal of the flower in itself contributes nothing to generation": Quoted in N. Gourlie, *The Prince of Botanists: Carl Linnaeus* (London: H. F. and G. Witherby, 1953), 30–31.

The opposite view was held by the spermists: J. Farley, *Gametes and Spores: Ideas about Sexual Reproduction 1750–1914* (Baltimore: Johns Hopkins University Press, 1982).

"Thou essence of dock, valerian and sage": L. L. Woodruff, "The Versatile Sir John Hill, M.D.," *American Naturalist* 60 (1926): 417–22.

"*Lucina sine Concubitu*, a letter humbly addressed": C. Emery, "'Sir' John Hill versus the Royal Society," *Isis* 13 (1942): 16-20.

The very rare Sahara Cypress *Cupressus dupreziana*, the twelfth-most endangered plant in the world: F. Abdoun and M. Beddiaf, "*Cupressus dupreziana* A. Camus: "Distribution, Decline and Regeneration on the Tassili n'Aijer, Central Sahara," *Comptes Rendus Biologies* 325 (2002): 617–627.

The pollen of *Cupressus dupreziana* was used to fertilize another species of *Cupressus*: C. Pichot, M. El Maataoui, S. Raddi, and P. Raddi, "Surrogate mother for endangered *Cupressus*," *Nature* 412 (2001): 39.

Other pollen grains are even more abnormal: M. El Maataoui, and C. Pichot, "Microsporogenesis in the Endangered Species *Cupressus dupreziana* A. Camus: Evidence for Meiotic Defects Yielding Unreduced and Abortive Pollen," *Planta* 213 (2001): 543–49.

Pushing the population nearer and nearer the brink of extinction: M. J. McKone and S. L. Halpern, "The Evolution of Androgenesis," *American Naturalist* 161 (2003): 641–56.

"Come away, come away, death / And in sad cypress let me be laid": Shakespeare, *Twelfth Night* (Collins edition), act 2, scene 4.

Rubus nessensis . . . appears to belong to a single apomictic clone: H. Nybom, "Biometry and DNA Fingerprinting Detect Limited Genetic Differentiation among Populations of the Apomictic Blackberry *Rubus nessensis* (Rosaceae)," *Nordic Journal of Botany* 18 (1998): 323–33.

I surveyed the results of several hundred studies of the genetic composition of plants: J. Silvertown, "The Evolutionary Maintenance of Sex: Evidence from the Ecological Distribution of Asexual Reproduction in Clonal Plants," *International Journal of Plant Sciences* 169 (2008): 157–68.

Reynoutria japonica is a good example. The entire alien population . . . consists of a single asexual clone!: M. L. Hollingsworth and J. P. Bailey, "Evidence for Massive Clonal Growth in the Invasive Weed *Fallopia japonica* (Japanese Knotweed)," *Botanical Journal of the Linnean Society* 133 (2000): 463–72.

The very first idea that was ever offered on the subject by Thomas Hunt Morgan: T. H. Morgan, *Heredity and Sex* (New York: Columbia University Press, 1913), 13–14.

Copies of beneficial genetic mutations are multiplied by natural selection and accumulate over the generations: G. Martin, S. P. Otto, and T. Lenormand, "Selection for Recombination in Structured Populations," *Genetics* 172 (2006): 593–609.

The entire species was in fact a single clone that the Romans brought to Britain two thousand years ago: L. Gil, P. Fuentes-Utrilla, A. Soto, M. T. Cervera, and C. Collada, "English Elm is a 2,000-year-old Roman Clone," *Nature* (2004): 431, 1053.

Genetic isolation also means that deleterious mutations can accumulate within a clonal lineage: J. de Visser and S. F. Elena, "The

Evolution of Sex: Empirical Insights into the Roles of Epistasis and Drift," *Nature Reviews Genetics* 8 (2007): 139–49.

Swamp loosestrife *Decodon verticillatus*: M. E. Dorken, K. J. Neville, and C. G. Eckert, "Evolutionary Vestigialization of Sex in a Clonal Plant: Selection versus Neutral Mutation in Geographically Peripheral Populations," *Proceedings of the Royal Society of London Series B-Biological Sciences* 271 (2004): 2375–80.

FURTHER READING

Few popular books tackle the basic question of why sex exists. I recommend Mark Ridley's *Mendel's Demon* (London: Phoenix, 2000). Not surprisingly, flowers receive much more attention, for example in Peter Bernhardt, *The Rose's Kiss: A Natural History of Flowers* (Chicago: University of Chicago Press, 2002).

Chapter Four } BEFORE THE SEED: POLLINATION

SOURCES

The mystery *The Naval Treaty*: From *The Naval Treaty* in A. Conan Doyle, *The Complete Sherlock Holmes Short Stories* (London: John Murray, 1928). Also see R. Milner, "Mystery of the Red Rose," *Natural History* 108 (1999): 36–39.

"There is weighty and abundant evidence": C. Darwin, *The Effects of Cross and Self-fertilization in the Vegetable Kingdom* (London: John Murray, 1876), chap. 1.

"In several flowers sent me by Mr. Bateman": C. Darwin, *The Various Contrivances by Which Orchids Are Fertilized by Insects*, 2nd ed. (London: John Murray, London, 1877), 162–63.

Not until the late 1990s was the moth actually observed in action: L. T. Wasserthal, "The Pollinators of the Malagasy Star Orchids *Angraecum sesquipedale*, *A. sororium*, and *A. compactum* and the

Evolution of Extremely Long Spurs by Pollinator Shift," *Botanica Acta* 110 (1997): 343–59.

Darwin's fascination with flower pollination was not a whimsical pursuit: D. Kohn, "The Miraculous Season," *Natural History* 114 (2005): 38–40.

Francis Darwin wrote that the book "not only encouraged . . .": F. Darwin, ed., *Life of Charles Darwin* (London: John Murray, 1902), 300.

He at first thought that the obvious experiment: C. Darwin, *The Effects of Cross and Self-fertilization in the Vegetable Kingdom* (London: John Murray, 1876).

A very general phenomenon known as inbreeding depression: B. C. Husband and D. W. Schemske, "Evolution of the Magnitude and Timing of Inbreeding Depression in Plants," *Evolution* 50 (1996): 54–70.

Four of Josiah Wedgewood II's seven children who reached adulthood: E. Healey, *Emma Darwin: The Inspirational Wife of a Genius* (London: Headline, 2001).

"My dread is hereditary ill-health": R. Keynes, *Annie's Box* (London: Fourth Estate, 2001), 208.

Asked the British parliament to consider adding a question to the national census of 1871: J. Moore, "Darwin Doubted His Own Family's 'fitness.'" *Natural History* 114 (2005): 45–46.

"The most wonderful case of fertilization ever published." The Darwin correspondence online database, letter 9395, dated 7 April 1874, Darwin to Hooker. http://darwin.lib.cam.ac.uk/perl/nav?pclass=calent;pkey=9395 (accessed 25 March 2007).

A female yucca moth has mouthparts with a unique tentacle-like structure: O. Pellmyr, "Yuccas, Yucca Moths, and Coevolution: A Review," *Annals of the Missouri Botanical Garden* 90 (2003): 35–55.

Four more examples of the phenomenon have come to light: O. Pellmyr, M. Balcazar-Lara, D. M. Althoff, K. A. Segraves, and J. Leebens-Mack, "Phylogeny and Life History Evolution of *Prodoxus* Yucca Moths (Lepidoptera: Prodoxidae)," *Systematic Entomology* 31 (2006): 1–20.

The ancestors of the Moraceae had wind-pollinated flowers: S. L. Datwyler and G. D. Weiblen, "On the Origin of the Fig: Phylogenetic Relationships of Moraceae from *ndhF* Sequences," *American Journal of Botany* 91 (2004): 767–77.

Pollination by the tiny wasps . . . originated at least 60 million years ago: N. Ronsted, G. D. Weiblen, J. M. Cook, N. Salamin, C. A. Machado, and V. Savolainen, "60 Million Years of Co-divergence in the Fig-Wasp Symbiosis," *Proceedings of the Royal Society B-Biological Sciences* 272 (2005): 2593–99.

The edible fig is not a fruit in the strict sense: M. Proctor and P. Yeo, *The Pollination of Flowers* (London: Collins, 1973).

Paternity testing of fig seeds: J. D. Nason, E. A. Herre, and J. L. Hamrick, "The Breeding Structure of a Tropical Keystone Plant Resource," *Nature* 391 (1998): 685–87.

FURTHER READING

A wide-ranging, accessible and thorough overview of pollination may be found in M. Proctor, P. Yeo, and A. Lack, *The Natural History of Pollination* (London: Collins, 1996). David Kohn's brief article, "The Miraculous Season," which appeared in *Natural History* 114 (2005): 38–40, gives a compelling account of how Charles Darwin spent the spring and summer following the publication of *On the Origin of Species* hidden away from all the fuss it created, quietly studying flowers and laying the foundations of a revolution in botany.

Chapter Five } ACCORDING TO THEIR OWN KINDS: INHERITANCE

SOURCES

"Here there were to be seen": H. Iltis, *Life of Mendel* (London: Allen and Unwin, 1932), 107.

Mendel, wasted many years of his life: R. M. Henig, *A Monk and*

Two Peas (London: Weidenfeld and Nicholson,2000).

Corn was the plant that earned the geneticist Barbara McClintock a Nobel Prize: N. C. Comfort, *The Tangled Field: Barbara McClintock's Search for the Patterns of Genetic Control* (Cambridge, MA: Harvard University Press, 2003).

It is estimated that 95% of the barley genome is just transposons: A. H. Schulman and R. Kalendar, "A Movable Feast: Diverse Retrotransposons and Their Contribution to Barley Genome Dynamics," *Cytogenetic and Genome Research* 110 (2005): 598–605.

FURTHER READING

A very readable recent biography of Mendel and the repercussions of his work is R. M Henig's *A Monk and Two Peas* (London: Weidenfeld and Nicholson, 2000). Barbara McClintock's research on corn is difficult to grasp in depth, but if you want to try, then R. N. Jones has written a well-illustrated and accessible account in "McClintock's Controlling Elements: The Full Story," *Cytogenetic and Genome Research* 109 (2005): 90–103. An engaging popular book on genetics is J. D. Ackerman's *Chance in the House of Fate: A Natural History of Heredity* (London: Bloomsbury, 2001).

Chapter Six } O ROSE, THOU ART SICK!: ENEMIES

SOURCES

Xanten, on the lower reaches of the Rhine, was struck by Divine Wrath: T. I. Williams, *Drugs from Plants* (London: Sigma Books, 1947).

At least 132 epidemics of ergotism were recorded in Europe between 591 and 1789: K. F. Kiple and K. C. Ornelas, eds., *The Cambridge World History of Food* (Cambridge: Cambridge University Press, 2000), 151.

Oliver Cromwell may have died of this form of ergot poisoning: M.

K. Matossian, *Poisons of the Past: Molds, Epidemics, and History* (New Haven: Yale University Press, 1989).

The evidence that the events in Salem . . . were caused: M. K. Mattossian, *Poisons of the Past: Molds, Epidemics, and History* (New Haven: Yale University Press, 1989).

The corn smut infects its host via the flowers: J. K. Pataky and M. A. Chandler, "Production of Huitlacoche, *Ustilago maydis*: Timing Inoculation and Controlling Pollination," *Mycologia* 95 (2003): 1261–70.

In the most widespread group of endophytes: C. L. Schardl, A. Leuchtmann, and M. J. Spiering, "Symbioses of Grasses with Seedborne Fungal Endophytes," *Annual Review of Plant Biology* 55 (2004): 315–40.

Eye of newt, and toe of frog: Shakespeare, *Macbeth* (Collins edition), act 4, scene 1, lines 14–15.

About six hundred . . . fungi are known to infect seeds: B. Spooner and P. Roberts, *Fungi*, 1st ed. (London: Harper Collins, 2005).

The insect performs a hormonal deception: S. Chiwocha, G. Rouault, S. Abrams, and P. von Aderkas, "Parasitism of Seed of Douglas Fir (*Pseudotsuga menziesii*) by the Seed Chalcid, *Megastigmus spermotrophus*, and Its Influence on Seed Hormone Physiology," *Sexual Plant Reproduction* 20 (2007): 19–25.

Red crossbills and lodgepole pines are eyeball-to-eyeball in a war over seeds: C. W. Benkman, T. L. Parchman, A. Favis, and A. M. Siepielski, "Reciprocal Selection Causes a Coevolutionary Arms Race between Crossbills and Lodgepole Pine," *American Naturalist* 162 (2003): 182–94.

FURTHER READING

A fascinating book on the role of ergotism in history is M. K. Matossian, *Poisons of the Past: Molds, Epidemics, and History* (New Haven: Yale University Press, 1989). For a general book about diseases of plants read D. Ingram and N. Robertson, *Plant Disease* (London: Harper Collins, 1999).

Chapter Seven } THE BIGGEST COCONUT I EVER SEE: SIZE

SOURCES

John Agard, "Palm Tree King," in *Flora Poetica*, ed. Sarah Maguire, *The Chatto Book of Botanical Verse* (London: Chatto & Windus, 2001), 205.

Peter Edwards and two colleagues: P. J. Edwards, J. Kollmann, and K. Fleischmann, "Life History Evolution in *Lodoicea maldivica* (Arecaceae)," *Nordic Journal of Botany* 22 (2002): 227–37.

A significant part of the variation in seed size is due to differences in growth form between species: A. T. Moles et al., "Factors That Shape Seed Mass Evolution," *Proceedings of the National Academy of Sciences of the United States of America* 102 (2005): 10540–44.

Seeds of palms as a group are over four hundred times bigger: A. T. Moles et al., "A Brief History of Seed Size," *Science* 307 (2005): 576–80.

FURTHER READING

A big (coffee-table-sized) book about big (and small) seeds with artistic photographs: R. Kessler and W. Stuppy, *Seeds: Time Capsules of Life* (London: Papadakis, 2006).

Chapter Eight } TEN THOUSAND ACORNS: NUMBER

SOURCES

Squirrels can discriminate between weevil-infested acorns: M. A. Steele, L. Z. HadjChikh, and J. Hazeltine, "Caching and Feeding Decisions by *Sciurus carolinensis*: Responses to Weevil-infested Acorns," *Journal of Mammalogy* 77 (1996), 305–14.

Certain individual oaks have evolved a countermeasure to this threat: A. B. McEuen and M. A. Steele, "Atypical Acorns Appear to Allow Seed Escape after Apical Notching by Squirrels," *American Midland Naturalist* 154 (2005): 450–58.

In former times humans ate them too: W. B. Logan, *Oak: The Frame of Civilization* (New York: W. W. Norton, 2005).

Acorns were also a staple for Native Americans: W. B. Logan, *Oak: The Frame of Civilization* (New York: W. W. Norton, 2005).

A study in forests in New York state showed: C. G. Jones, R. S. Ostfeld, M. P. Richard, E. M. Schauber, and J. O. Wolff, "Chain Reactions Linking Acorns to Gypsy Moth Outbreaks and Lyme Disease risk," *Science* 279 (1998): 1023–26.

The gypsy moth is an introduced pest: U.S. Forest Service Web site on the gypsy moth problem, http://www.fs.fed.us/ne/morgantown/4557/gmoth/.

"All is for the best in the best of all possible worlds": Voltaire, *Candide.*

New England has more forest cover now: D. R. Foster, "Thoreau's Country: A Historical-ecological Perspective on Conservation in the New England Landscape," *Journal of Biogeography* 29 (2002): 1537–55.

"On some streets, it was just one house after another": Dr. Allen Steere, quoted in J. A. Edley, *Bull's Eye: Unravelling the Medical Mystery of Lyme Disease*, 2nd ed. (New Haven: Yale University Press, 2004), 40.

The deer population has increased dramatically: D. R. Foster, G. Motzkin, D. Bernardos, and J. Cardoza, "Wildlife Dynamics in the Changing New England Landscape," *Journal of Biogeography* 29 (2002): 1337–57.

Synchrony over distances of up to 2,500 thousand kilometers: W. D. Koenig and J. M. H. Knops, "Scale of Mast-seeding and Tree-ring Growth," *Nature* 396 (1998): 225–26.

Eruption of seed-eating birds: W. D. Koenig and J. M. H. Knops, "Seed-crop Size and Eruptions of North American Boreal Seed-eating Birds," *Journal of Animal Ecology* 70 (2001): 609–20.

Why do oaks and so many other forest trees . . . vary their annual seed production so drastically from year to year?: J. Silvertown, "The Evolutionary Ecology of Mast Seeding in Trees," *Biological Journal of the Linnean Society* 14 (1980): 235–50.

Dipterocarp trees . . . mast years: P. S. Ashton, T. J. Givnish, and S. Appanah, "Staggered Flowering in the Dipterocarpaceae: New Insights into Floral Induction and the Evolution of Mast Fruiting in the Aseasonal Tropics," *American Naturalist* 132 (1998): 44–66.

Masting is much more extreme than weather variation: W. D. Koenig and J. M. H. Knops, "Patterns of Annual Seed Production by Northern Hemisphere Trees: A Global Perspective," *American Naturalist* 155 (2000): 59–69.

It has been calculated that by destroying potential breeding places for the pecan weevil: M. Harris and C. S. Chung, "Masting Enhancement Makes Pecan Nut Casebearer Pecans Ally against Pecan Weevil," *Journal of Economic Entomology* 91 (1998): 1005–10.

For fruit trees it would be counterproductive to satiate their dispersal agents: J. Silvertown, "The Evolutionary Ecology of Mast Seeding in Trees," *Biological Journal of the Linnean Society* 14 (1980): 235–50.

The lining of the shell contains an oil: F. Rosengarten Jr., *The Book of Edible Nuts* (Mineola, NY: Dover Publications, 2004).

FURTHER READING

The bounty of acorns to be had from oaks and how Native Americans used them is described by William Bryant Logan in *Oak: The Frame of Civilization* (New York: Norton, 2005). The American chestnut, now sadly destroyed by chestnut blight, produced nuts so abundantly that it once supported a whole rural economy. This is wonderfully described by Susan Freinkel in *American Chestnut* (Berkeley: University of California Press, 2007). Jonathan Edlow readably unravels the medical mystery of Lyme disease in *Bull's Eye: Unravelling the Medical Mystery of Lyme Disease*, 2nd ed. (New Haven: Yale University Press, 2004).

Chapter Nine } LUSCIOUS CLUSTERS OF THE VINE: FRUIT

SOURCES

Spread so successfully that they have become serious weeds: J. Silvertown, *Demons in Eden: The Paradox of Plant Diversity* (Chicago: University of Chicago Press, 2005).

Hamilton and May's model of dispersal: W. D. Hamilton and R. M. May, "Dispersal in Stable Habitats," *Nature* 269 (1977): 578–81.

More and more cases of directed dispersal are now coming to light: D. G. Wenny, "Advantages of Seed Dispersal: A Re-evaluation of Directed Dispersal," *Evolutionary Ecology Research* 3 (2001): 51–74.

A mistle thrush will stoutly defend a tree containing mistletoe from other birds during the berrying season: B. Snow and D. Snow, *Birds and Berries: A Study of an Ecological Interaction* (Calton, UK: T. and A. D. Poyser,1988).

Perches on branches of just the right diameter for the establishment of mistletoe seedlings: D. G. Wenny, "Advantages of Seed Dispersal: A Re-evaluation of Directed Dispersal," *Evolutionary Ecology Research* 3 (2001): 51–74:

The first fleshy upholstery around seeds was not technically a fruit . . . but belonged to early seed plants: B. H. Tiffney, "Vertebrate Dispersal of Seed Plants through Time," *Annual Review of Ecology Evolution and Systematics* 35 (2004): 1–29.

Fruit makes up a large part of the diet: K. Milton, "Ferment in the Family Tree: Does a Frugivorous Dietary Heritage Influence Contemporary Patterns of Human Ethanol Use?" *Integrative and Comparative Biology* 44 (2004): 304–14.

Dogs and horses have no cones that peak in the mid (green) part of the spectrum: T. L. Dawson, "Colour and Colour Vision of Creatures Great and Small," *Coloration Technology* 122 (2006): 61–73.

In a study that compared human subjects with two-color . . . with normal subjects: M. J. Morgan, A. Adam, and J. D. Mollon, "Dichromates Detect Color-camouflaged Objects That Are Not Detected by Trichromates," *Proceedings of the Royal Society of London Series B-Biological Sciences* 248 (1992): 291–95.

Lab studies with nonhuman primates obtained the same result: A. Saito et al., "Advantage of Dichromats over Trichromats in Discrimination of Color-camouflaged Stimuli in Nonhuman Primates," *American Journal of Primatology* 67 (2005): 425–36.

The fruits that birds are attracted to are mainly red or black in color: M. F. Willson and C. J. Whelan, "The Evolution of Fruit Color in Fleshy-fruited Plants," *American Naturalist* 136 (1990): 790–809.

Are UV-reflecting and birds use this wavelength in foraging for them: H. Siitari et al., "Ultraviolet Reflection of Berries Attracts Foraging Birds: A Laboratory Study with Redwings (*Turdus iliacus*) and Bilberries (*Vaccinium myrtillus*)," *Proceedings of the Royal Society of London Series B-Biological Sciences* 266 (1999): 2125–29; D. L. Altshuler, "Ultraviolet Reflectance in Fruits: Ambient Light Composition and Fruit Removal in a Tropical Forest," *Evolutionary Ecology Research* 3 (2001): 767–78.

Studies of this kind have been verified by behavioral experiments: A. C. Smith, H. M. Buchanan-Smith, A. K. Surridge, D. Osorio, and N. I. Mundy, "The Effect of Colour Vision Status on the Detection and Selection of Fruits by Tamarins (*Saguinus* spp.)," *Journal of Experimental Biology* 206 (2003): 3159–76.

Red fruit show up much better: D. Osorio et al., "Detection of Fruit and the Selection of Primate Visual Pigments for Color Vision," *American Naturalist* 164 (2004): 696–708.

Fruit ripeness . . . is easier to judge: P. Riba-Hernandez et al., "Sugar Concentration of Fruits and Their Detection via Color in the Central American Spider Monkey (*Ateles geoffroyi*)," *American Journal of Primatology* 67 (2005): 411–23.

Humans with defective color vision have difficulty finding fruit among foliage: B. C. Regan et al., "Fruits, Foliage and the Evolution of Primate Colour Vision," *Philosophical Transactions of the Royal Society of London Series B-Biological Sciences* 356 (2001): 229–83.

Alu is found in suspicious association with the opsin gene: K. S. Dulai et al., "The Evolution of Trichromatic Color Vision by Opsin Gene Duplication in New World and Old World Primates," *Genome Research* 9 (1999): 629–38.

FURTHER READING

An up-to-date account of bird vision that is well worth reading is given in the article "What Birds See," by T. H. Goldsmith, *Scientific American* 295 (2006): 68–75. *The Story of the Apple*, by Barry E. Juniper and David J. Mabberley (Portland: Timber Press, 2006), does what it says on the tin and describes the evolution of this particular fleshy fruit in the wilds of Central Asia as well as its long association with humans.

Chapter Ten } WINGED SEEDS: DISPERSAL

SOURCES

Inspired the Austrian Igo Etrich: www.century-of-flight.freeola.com/Aviation%20history/flying%20wings/Early%20Flying%20Wings.htm (accessed 20.04.07).

The other notable example occurs, like *Alsomitra*, in a tropical forest vine: R. Kessler and W. Stuppy, *Seeds: Time Capsules of Life* (London: Papadakis, 2006), 97.

Brazilian zebra wood tree (*Centrolobium robustum*): R. Kessler and W. Stuppy, *Seeds: Time Capsules of Life* (London: Papadakis, 2006), 94.

Until wind of sufficient strength is able to whisk them away: P. Schippers and E. Jongejans, "Release Thresholds Strongly Determine the Range of Seed Dispersal by Wind," *Ecological Modelling* 185 (2005): 93–103.

Orangutans use their weight: S. K. S. Thorpe, R. H. Crompton, and R. M. Alexander, "Orangutans Use Compliant Branches to Lower the Energetic Cost of Locomotion," *Biology Letters* 3 (2007): 253–56.

Henry Horn solved this problem in typically ingenious fashion: H. S. Horn, "Eddies at the Gates," *Nature* 436 (2005): 179.

In the frozen North lodgepole pine is captured in the act of colonization: L. C. Cwynar and G. M. MacDonald, "Geographical Variation of Lodgepole Pine in Relation to Population History," *American Naturalist* 129(1987): 463–69.

FURTHER READING

As you might expect, more has been written about flying aircraft than flying seeds. The story of the Wright brothers is succinctly told in a book illustrated with period photographs by Fred Culick and Spencer Dunmore: *On Great White Wings: The Wright Brothers and the Race for Flight* (New York: Hyperion, 2001).

Chapter Eleven } CIRCUMSTANCE UNKNOWN: FATE

SOURCES

"The concrete highway was edged": John Steinbeck, *The Grapes of Wrath* (London: William Heinemann, 1939), chap. 3.

A simple bristle on a grass seed could pilot it to a vertical landing: M. H. Peart, "Experiments on the Biological Significance of the Morphology of Seed-dispersal Units in Grasses," *Journal of Ecology* 67 (1979): 843–63.

Three thousand species belonging to eighty different plant families: I. Giladi, "Choosing Benefits or Partners: A Review of the Evidence for the Evolution of Myrmecochory," *Oikos* 112 (2006): 481–92.

Stick insects have evolved eggs that mimic the appearance of seeds:

L. Hughes and M. Westoby, "Capitula on Stick Insect Eggs and Elaiosomes on Seeds: Convergent Adaptations for Burial by Ants," *Functional Ecology* 6 (1992): 642–48.

The chemical composition of elaiosomes: L. Hughes, M. Westoby, and E. Jurado, "Convergence of Elaiosomes and Insect Prey: Evidence from Ant Foraging Behaviour and Fatty Acid Composition," *Functional Ecology* 8 (1994): 358–65.

Fleshy fruit that contain seeds furnished with an elaiosome: S. B. Vander Wall and W. S. Longland, "Diplochory: Are Two Seed Dispersers Better Than One?" *Trends in Ecology and Evolution* 19 (2004): 155–61.

The very earth itself is a granary: H. D. Thoreau, *Faith in a Seed* (Washington, DC: Island Press, 1993).

A two-thousand-year-old date seed: S. Sallon, E. Solowey, Y. Cohen, R. Korchinsky, M. Egli, I. Woodhatch, O. Simchoni, M. Kislev, "Germination, genetics, and growth of an ancient date seed," *Science* 320 (2008): 1464.

These seeds have never been successfully germinated: *The Encyclopedia of Seeds: Science, Technology, and Uses*, ed. M. Black, J. Bewley, and P. Halmer (Cambridge, MA: CABI, 2006).

Seeds among documents belonging to a Dutch merchant: M. Daws, "Seed Survives for 200 Years," *Kew Scientist* 31 (2007): 2.

"You have in your drawer since Christmas Day": Lawrence D. Hills, reproduced in P. Loewer, *Seeds: The Definitive Guide to Growing, History, and Lore* (New York: John Wiley and Sons, 1995), 256.

Seeds in the soil, which may reach tens of thousands per square meter: K. Thompson, J. Bakker, and R. Bekker, *The Soil Seed Banks of North West Europe* (Cambridge: Cambridge University Press, 1996).

FURTHER READING

The standard textbook on this and related ecological topics is by Mike Fenner and Ken Thompson, *The Ecology of Seeds* (Cambridge: Cambridge University Press, 2005).

Chapter Twelve } FIERCE ENERGY: GERMINATION

SOURCES

Such is the force exerted by a germinating seed: http://waynesword.palomar.edu/pljuly96.htm (accessed 3 July 2007).

South African klapperbossie *Blepharis mitrata*: R. Kessler and W. Stuppy, *Seeds: Time Capsules of Life* (London: Papadakis, 2006), 111.

Germinate in spring if they have experienced fluctuating temperatures: K. Thompson and J. P. Grime, "A Comparative Study of Germination Responses to Diurnally-fluctuating Temperatures," *Journal of Applied Ecology* 20 (1983): 141–56.

Desmodium species exude two substances: Z. R. Khan, A. Hassanali, W. Overholt, T. M. Khamis, A. M. Hooper, J. A. Pickett, L. J. Wadhams, and C. M. Woodcock, "Control of Witchweed *Striga hermonthica* by Intercropping with *Desmodium* spp., and the Mechanism Defined as Allelopathic," *Journal of Chemical Ecology* 28 (2002): 1871–85.

FURTHER READING

If you want to go further into this subject, then you should consult the monumental work by Carol Baskin and Jerry Baskin, *Seeds: Ecology, Biogeography, and Evolution of Dormancy and Germination* (San Diego: Academic Press, 1998).

Chapter Thirteen } SORROW'S MYSTERIES: POISONS

SOURCES

Pythagoras forbade his followers to eat beans: K. Albala, *Beans: A History* (Oxford: Berg, 2007).

Seeds of the neem tree from Burma: A. Beattie and P. R. Ehrlich, *Wild Solutions* (New Haven: Yale University Press, 2001).

Ten percent of adult deaths among Chamorros were due to lytico-bodig: Oliver Sacks, *The Island of the Colour Blind* (London: Picador, 1996), 19.

Flying fox tissue contained concentrations of BMAA that were a thousand times greater than in fadang flour: P. A. Cox, S. A. Banack, and S. J. Murch, "Biomagnification of Cyanobacterial Neurotoxins and Neurodegenerative Disease among the Chamorro People of Guam," *Proceedings of the National Academy of Sciences of the United States of America* 100 (2003): 13380–83. S. J. Murch, P. A. Cox, and S. A. Banack, "A Mechanism for Slow Release of Biomagnified Cyanobacterial Neurotoxins and Neurodegenerative Disease in Guam," *Proceedings of the National Academy of Sciences of the United States of America* 101 (2004): 12228–31.

Extinction of one species of flying fox and the near extinction of another: C. S. Monson, S. A. Banack, and P. A. Cox, "Conservation Implications of Chamorro Consumption of Flying Foxes as a Possible Cause of Amyotrophic Lateral Sclerosis–Parkinsonism Dementia Complex in Guam," *Conservation Biology* 17 (2003): 678–86.

BMAA can accumulate in a hidden reservoir, bound to proteins: S. J. Murch, P. A. Cox, and S. A. Banack, "A Mechanism for Slow Release of Biomagnified Cyanobacterial Neurotoxins and Neurodegenerative Disease in Guam," *Proceedings of the National Academy of Sciences of the United States of America* 101 (2004): 12228–31.

BMAA from some other environmental source, as cyanobacteria occur in many aquatic and terrestrial ecosystems: P. A. Cox et al. "Diverse Taxa of Cyanobacteria Produce Beta-N-methyl-amino-L-alanine, a Neurotoxic Amino Acid," *Proceedings of the National Academy of Sciences of the United States of America* 102 (2005): 5074–78.

FURTHER READING

The beginnings of the lytico bodig story are recounted in Oliver Sacks's book *The Island of the Colour Blind* (London: Picador, 1996).

Chapter Fourteen } AH, SUN-FLOWER!: OIL

SOURCES

Countless poets have exercised their license to imagine otherwise: C. B. Heiser, *The Sunflower* (Norman: University of Oklahoma Press, 1976), chap. 3.

Wesley Powell described seed harvesting and preparation: C. B. Heiser, *The Sunflower* (Norman: University of Oklahoma Press, 1976), 30–31.

"As man began to use the plant": C. B. Heiser, *The Sunflower* (Norman: University of Oklahoma Press, 1976), 81–82.

Eastern North America's claim to have the earliest domesticated sunflower was trumped by an earlier find in Tabasco: D. L. Lentz, M. E. D. Pohl, K. O. Pope, and A. R. Wyatt, "Prehistoric Sunflower (*Helianthus annuus* L.) Domestication in Mexico," *Economic Botany* 55 (2001): 370–76.

In 2004 a study of this kind compared the DNA of sunflower cultivars with that of wild populations: A. V. Harter, K. A. Gardner, D. Falush, D. L. Lentz, R. A. Bye, and L. H. Rieseberg, "Origin of Extant Domesticated Sunflowers in Eastern North America," *Nature* 430 (2004): 201–5.

In the early nineteenth century the Russian Orthodox Church: C. B. Heiser, *The Sunflower* (Norman: University of Oklahoma Press, 1976).

Trade in seed oils is worth more than $61 billion a year: D. J. Murphy, "The Biogenesis and Functions of Lipid Bodies in Animals, Plants, and Microorganisms," *Progress in Lipid Research* 40 (2001): 325–438.

To tell the time, several nuts: http://en.wikipedia.org/wiki/Candlenut (accessed 12.02.2006).

Rainforest in Southeast Asia is being cleared for oil palm plantations: George Monbiot, *Worse than Fossil Fuel.* http://www.

monbiot.com/archives/2005/12/06/worse-than-fossil-fuel/ (accessed 12.02.2006).

Plant and animal oils and fats: C. M. Pond, *The Fats of Life* (Cambridge: Cambridge University Press, 1998).

All triacylglycerols a plant produces are first synthesized in saturated form: C. R. Linder, "Adaptive Evolution of Seed Oils in Plants: Accounting for the Biogeographic Distribution of Saturated and Unsaturated Fatty Acids in Seed Oils," *American Naturalist* 156 (2000): 442–58.

This pattern is so general: C. R. Linder, "Adaptive Evolution of Seed Oils in Plants: Accounting for the Biogeographic Distribution of Saturated and Unsaturated Fatty Acids in Seed Oils," *American Naturalist* 156 (2000): 442–58.

Oils and fats have the unusual property of behaving like liquids: Caroline M. Pond, *The Fats of Life* (Cambridge: Cambridge University Press, 1998).

Castor oil and derivatives: www.killerplants.com/plants-that-changed-history/20040210.asp (accessed 04.07.07).

FURTHER READING

The definitive book on sunflowers, although now a little out of date is Charles Heiser's *The Sunflower* (Norman: University of Oklahoma Press, 1976). Although sunflower seeds are not technically nuts, they feature in F. J. Rosengarten's *The Book of Edible Nuts* (Mineola, NY: Dover, 2004). If you are at all interested in seeds as food from any point of view, then *The Book of Edible Nuts* is a must. *The Fats of Life* by Caroline M. Pond (Cambridge: Cambridge University Press, 1998) is a gold mine of fascinating information about lipids, including those found in plants.

Chapter Fifteen } JOHN BARLEYCORN: BEER

SOURCES

"The nations of the West also have their own intoxicant": Quoted in H. McGee, *McGee on Food and Cooking* (London: Hodder and Stoughton,2004), 741.

Pliny was less disparaging of barley as food: Pliny the Elder, *The Natural History*, ed. H. T. Riley and J. Bostock, bk. 18, chap. 14; http://www.perseus.tufts.edu/cgi-bin/ptext?lookup=Plin.+Nat.+18.14 (accessed 15 February 2006).

Barley was important in the transition from hunter-gathering to settled agriculture: B. D. Smith, *The Emergence of Agriculture* (New York: Scientific American Library, 1998).

Two genes control whether barley shatters or not: G. C. Hillman and M. S. Davies, "Domestication Rates in Wild-Type Wheats and Barley under Primitive Cultivation," *Biological Journal of the Linnean Society* 39, no. 1(1990): 39–78.

Domestication wrought other important changes upon wild barley in the Fertile Crescent: D. Zohary and M. Hopf, *Domestication of Plants in the Old World* (Oxford: Oxford University Press, 2000).

To investigate the origins of wild yeast, see Justin Fay and Joseph Benavides: J. C. Fay and J. A. Benavides, "Evidence for Domesticated and Wild Populations of *Saccharomyces cerevisiae*," *PLoS Genetics* 1 (2005): 66–71.

The earliest evidence of fermented drinks so far uncovered is from China: P. E. McGovern et al., "Fermented Beverages of Pre- and Proto-historic China," *Proceedings of the National Academy of Sciences of the United States of America* 101 (2004): 17593–98.

The advantage to ethanol production is that it poisons other microbes: D. Janzen, "Why Fruits Rot, Seeds Mold, and Meat Spoils," *American Naturalist* 111 (1980): 691–713.

Studies of molecular evolution: J. M. Thomson, E. A. Gaucher, M. F. Burgan, D. W. De Kee, T. Li, J. P. Aris, and S. A. Benner,

"Resurrecting Ancestral Alcohol Dehydrogenases from Yeast," *Nature Genetics* 37 (2005): 630–35.

FURTHER READING

For an excellent account of how farming began, I recommend *The Emergence of Agriculture*, by B. D. Smith (New York: Scientific American Library, 1998). Plant-by-plant details of how barley and other Old World crops were domesticated can be found in the third edition of *Domestication of Plants in the Old World*, by D. Zohary and M. Hopf (Oxford: Oxford University Press, 2000). A popular account of beer making is *Beer: Tap into the Art and Science of Brewing*, by C. W. Bamforth (Oxford: Oxford University Press, 2003).

Chapter Sixteen } REALM OF ILLUSION: COFFEE

SOURCES

As germination begins, the embryo coffee plant . . . : T. W. Baumann, in *Espresso Coffee: The Science of Quality*, ed. A. Illy and R. Viani (San Diego: Elsevier, 2005), 55–67.

C. arabica beans contain less than 1.5 percent caffeine: E. Illy, "The Complexity of Coffee," *Scientific American* 286, no. 6 (2002).

"*Arabica* coffee's crude, boorish, sour, uncivilized black-hearted cousin": A. Wild, *Black Gold: A Dark History of Coffee* (London: Harper Perennial, 2005).

Ethiopians invaded Yemen and ruled there for a brief period: M. Pendergast, *Uncommon Grounds: The History of Coffee and How It Transformed Our World* (New York: Texere, 2001).

"Women's petition against Coffee": A. Wild, *Black Gold: A Dark History of Coffee* (London: Harper Perennial, 2005), 91.

"As soon as coffee is in your stomach . . ." Honoré de Balzac, quoted at www.cocoajava.com/java_quotes.html (accessed 9 July 2007).

Caffeine blocks the action of a substance called adenosine: J. W. Daly and B. B. Fredholm, "Mechanisms of Action of Caffeine on the Nervous System," in *Coffee, Tea, Chocolate, and the Brain*, ed. A. Nehlig (Boca Raton: CRC Press, 2004).

Adenosine accumulates during waking hours: J. Snel, Z. Tieges, and M. M. Lorist, "Effects of Caffeine on Sleep and Wakefulness: An Update," in *Coffee, Tea, Chocolate, and the Brain*, ed. A. Nehlig (Boca Raton: CRC Press, 2004).

"Increases arousal, vigilance": M. Casas, J. A. Ramos-Quiroga, G. Prat, and A. Qureshi, "Effects of Coffee and Caffeine on Mood and Mood Disorders," in *Coffee, Tea, Chocolate, and the Brain*, ed. A. Nehlig (Boca Raton: CRC Press, 2004).

Genealogy of coffee varieties recorded in their DNA: F. Anthony et al., "The Origin of Cultivated *Coffea arabica* L. Varieties Revealed by AFLP and SSR Markers," *Theoretical and Applied Genetics* 104 (2002): 894–900.

Many important commercial and scientific institutions were founded in coffee houses: M. Ellis, *The Coffee House: A Cultural History* (London: Weidenfeld & Nicholson, 2004), 304.

"Rally, Mohawks—bring out your axes!": A. Wild, *Black Gold: A Dark History of Coffee* (London: Harper Perennial, 2005), 134.

Ludwig ascribed his father's early death to caffeine-poisoning: H. E. Jacob, *The Saga of Coffee* (London: Allen and Unwin, 1935).

A coffee plant in Ethiopia that produces beans that are naturally lacking in caffeine: M. B. Silvarolla, P. Mazzafera, and L. C. Fazuoli, "A Naturally Decaffeinated Arabica Coffee," *Nature* 429 (2004): 826.

FURTHER READING

There are lots of books about coffee, and when you've read a few they all begin to seem alike. The Jamaican Blue Mountain of coffee books in my opinion is *Uncommon Grounds: The History of Coffee and How It Transformed Our World*, by Mark Pendergrast (New York: Texere, 2001). For all the technical detail you could want about coffee making, served at double espresso strength, try *Espresso Coffee: The Science of Quality*, ed. A. Illy and R. Viani (San Diego: Elsevier, 1995).

Chapter Seventeen } NOURISHMENT & INSPIRATION: GASTRONOMY

SOURCES

Sylvester Graham, who believed that its wholegrain recipe would help weaken sexual desire: K. F. Kiple and K. C. Ornelas, eds., *The Cambridge World History of Food* (Cambridge: University of Cambridge Press, 2000).

Not so harmless was the "macrobiotic" diet: K. F. Kiple and K. C. Ornelas, eds., *The Cambridge World History of Food* (Cambridge: University of Cambridge Press, 2000), 1565.

People who include significant amounts of whole grains in their diet have a substantially lower risk of cardiovascular disease: www.sciencedaily.com/releases/2007/05/070509161030.htm (accessed 4 July 2007).

Many health benefits, including lowering the risk of coronary heart disease: Harvard School of Public Health: http://www.hsph.harvard.edu/nutritionsource/fiber.html (accessed 19 August 2007).

The World Porridge-Making Championships: www.goldenspurtle.com (accessed 12 January 2008).

The full recipe is available on the Internet: www.bbc.co.uk/food/recipes/database/snailporridge_74858.shtml (accessed 12 January 2008).

Hindus who had been quite healthy on a vegan diet in their native India: K. F. Kiple and K. C. Ornelas, eds., *The Cambridge World History of Food* (Cambridge: University of Cambridge Press, 2000), 1567.

Experiments of this kind have found: D. Benton, "The Biology and

Psychology of Chocolate Craving," *Coffee, Tea, Chocolate, and the Brain*, ed. A. Nehlig (Boca Raton: CRC Press, 2004).

Berries of the West African tree *Synsepalum dulcificum*: L. Beidler and K. Kurihara, "Taste-Modifying Protein from Miracle Fruit," *Science* 161 (September 1968); McVicar Cannon, *The Old Sweet Lime Trick*, http://quisqualis.com/mirfrtdmc1a.html (accessed 4 July 2007).

The explanation for the change lies in the retina of the human eye: S. Kreft and M. Kreft, "Physicochemical and Physiological Basis of Dichromatic Colour," *Naturewissenschaften* 94 (2007): 935–39.

The Fat Duck, rated the world's best restaurant: www.fatduck.co.uk/ (accessed 18 August 2007).

Heston Blumenthal at the Fat Duck: www.fatduck.co.uk/ (accessed 18 August 2007).

The Maillard reaction: C. Billaud and J. Adrian, "Louis-Camille Maillard, 1878–1936," *Food Reviews International* 19 (2003): 345–74; P. A. Finot, "Historical Perspective of the Maillard Reaction in Food Science," *Annals of the New York Academy of Sciences* 1043 (2005): 1–8.

"I have been popping corn to-night, which is only a more rapid blossoming of the seed under a greater than July heat": Harold McGee, *McGee on Food and Cooking* (London: Hodder and Stoughton, 2004).

Kernels recovered from archaeological deposits that are thousands of years old can be popped: Betty Fussell, *The Story of Corn* (Albuquerque: University of New Mexico Press, 2004).

FURTHER READING

If you are interested in the science of food and cooking, you have to get hold of a copy of the incomparable *McGee on Food and Cooking*, by Harold McGee (London: Hodder and Stoughton, London, 2004). It's a wonderful amalgam of science, history and gastronomy. K. F. Kiple and K. C. Ornelas, eds., *The Cambridge World History of Food* (Cambridge: University of Cambridge Press, 2000) is another indispensable source which contains more than just a history of food and which in two volumes covers just about everything you can imagine wanting to know. There are lots of books about individual ingredients. Two that are particularly relevant here and which you should enjoy are Ken Albala's *Beans: A History* (Oxford: Berg, 2007); and Betty Fussell's *The Story of Corn* (Albuquerque: University of New Mexico Press, Albuquerque 2004).

Index